Management Dilemmas

The Theory of Constraints Approach to Problem Identification and Solutions

by Eli Schragenheim

ADVANCE PRAISE...

Learning to apply the Theory of Constraints has just gotten more exciting with this wonderful book! It is the book I've been looking for to encourage my students to become more involved in their education. *Management Dilemmas* has found a home in my classes.

—Richard E. Peschke, Ph.D.
Associate Professor of Management
Moorhead State University

This book is excellent! *Management Dilemmas* explores territory not well understood, and that is how TOC is applied to organizational analysis and strategy formulation. It also clearly shows how by focusing on one or two things, you can have a dramatic impact on the performance of an organization.

—Mark Woeppel
Rosewood Equipment Company

Eli Schragenheim, through his unique socratic writing style and real-life case study examples, manages to create a fascinating journey into some of the most challenging management dilemmas and offers valuable insight and focused direction to their root causes and solutions.

—Matti Herzberg
Chesapeake Consulting Inc.
(Formerly an associate of the Goldratt Institute)

A visionary practical book for every manager and consultant who likes to make real impact upon their organization. *Management Dilemmas* is also ideal for management workshops. I'm already using it in mine.

—Avner Passal, Ph.D.
Managing director, A.R Motive, Israel

Eli Schragenheim's stories are written in a way that makes us think, so give yourself the opportunity to do just that! As you read each story, try to analyze the case and derive your own solutions. Then, read his analyses carefully, and take in the powerful TOC focusing process that he illustrates so well.

—Lisa Scheinkopf
Chesapeake Consulting Inc.
Chairman, APICS Constraints Management SIG

The St. Lucie Press/APICS Series on Constraints Management

Series Advisors

Dr. James F. Cox, III	Thomas B. McMullen, Jr.
University of Georgia	*McMullen Associates*
Athens, Georgia	*Weston, Massachusetts*

Titles in the Series

Introduction to the Theory of Constraints (TOC) Management System
by Thomas B. McMullen, Jr.

Securing the Future: Strategies for Exponential Growth Using the Theory of Constraints
by Gerald I. Kendall

Project Management in the Fast Lane: Applying the Theory of Constraints
by Robert C. Newbold

The Constraints Management Handbook
by James F. Cox, III and Michael S. Spencer

Thinking for a Change: Putting the TOC Thinking Processes to Use
by Lisa J. Scheinkopf

Management Dilemmas: The Theory of Constraints Approach to Problem Identification and Solutions
by Eli Schragenheim

Management Dilemmas

Dilemmas

The Theory of Constraints
Approach to
Problem Identification
and Solutions

Eli Schragenheim

The St. Lucie Press/APICS Series on Constraints Management

S^t_L

St. Lucie Press

Boca Raton London   New York   Washington, D.C.

Library of Congress Cataloging-in-Publication Data

Schragenheim, Eli.
    Management dilemmas : the theory of constraints approach to problem identification
and solutions / Eli Schragenheim.
       p. cm.
    Includes bibliographical references and index.
    ISBN 1-57444-222-8 (alk. paper)
    1. Theory of constraints (Management)   2. Problem solving. I. Title
HD69.T46S37   1998
658.5--dc21

               98-41213
               CIP

Contents

Preface vii

The Author xi

About APICS xiii

Foreword xv

1   Basics of TOC and How to Analyze a Variety of Cases 1

2   When Things Are Going Well, Why Change? 33

3   Let There Be Light 47

4   A Laboratory for the Repair of Communication Equipment 65

5   We Need That Department 77

6   Where Is My Personal Buffer? 89

7   The Perspective of an Organizational Change 103

8   The Profit that Came from the Wrong Product 121

9   Crocodile Eyes: The Failure of a Great Project 131

10   Missing Information 147

11   Planning the Next Season and How to be Supported
       by the Information System 161

12   A Crisis at The Small News 177

13   Success Can Be a Problem, Too 189

Epilogue 207

Bibliography 209

Preface

The purpose of this book is to provoke thinking about management in general and about system constraints and core problems in particular. Although this books presents and highlights a particular management philosophy, I believe the way I've chosen to present the arguments lends itself to many other alternative approaches.

What do we expect from a "management philosophy"? I assume it should guide real world managers to make better decisions, meaning to take a course of action that helps an organization as a whole to better achieve its goal. The problem is that even after the fact, it is not simple to relate the actual results with a particular decision, as so many other actions and incidents have taken place in between. A substantial part of this book is to try to relate cause-and-effect relationships to the happenings in a certain situation. Another part is to speculate what will happen if a certain course of action takes place.

To achieve the purpose of provoking thoughts about cause and effects in a human-based system, this book presents a series of stories that describe problematic situations in a variety of organizations. The stories do not tell us all that has happened. They do not reach the natural end of a regular story. Most of them finish at the point where a question or a dilemma is raised. In this way, I as the author, cannot dictate my own theory. I just tell the story up to the point where some guidance or analysis is needed to shed a new light so the question can be answered and the dilemma can be resolved.

I believe that creating a virtual experience through stories, case studies, or computerized simulations is a great way to learn. For years I have developed computerized case studies that challenge the users to "manage" a virtual organization, thus providing situations in which critical questions about cause and effects are raised.

Whenever I was not able to present computerized simulations as a means to provoke thinking about the pressing issues of a particular organization, I have written a story that is close enough to the particular management group but also is detached from the daily details in order to refrain from emotions that block the ability to think afresh.

In this book I use some of the stories I have used in the past, but the scope is now much larger and much more generic. I have intentionally included a variety of organizations and situations to demonstrate that a structured methodology can be used to analyze the cause-and-effect relationship and come up with an improved understanding of why the results are what they are and what is likely to happen next. As the stories are merely stories on paper, I fully realize that there is no proof that the suggested method works in reality. There is no proof that any analysis is right. There is no proof that what is predicted to happen will actually happen. There is even no proof that the facts in the story are real facts, as the story itself is fiction and may contain impossible or unlikely incidents. Of course, I did my best to write fiction that according to my own experience is perfectly possible. Some of the stories are based on real situations, but I intentionally changed the original facts, so I can prove nothing except common sense. And common sense is open to debate.

In general, I feel much less a teacher than a facilitator for self-learning about an exciting and different way of thinking about business and management. The option of the reader to disagree with my analysis of a case or the method I chose for analysis only makes the book more thought provoking.

I consider myself a pupil of Dr. Eli Goldratt, founder of the Theory of Constraints (TOC). The seven years during which I worked very closely with Dr. Goldratt had an enormous positive impact on my thinking. Over the last five years I have tried to create my own personal interpretation of what I learned during those seven years. I will try to communicate in this book some of my learning and generic thinking guidelines.

The power of TOC lies in its way of simplifying complex and confusing situations by concentrating on the few most significant factors that are responsible for the performance of the organization as a whole. Being able to immediately find the few core problems behind the myriad "hurts" is much easier said than done, but I believe TOC provides a way for us to train ourselves to do that. I also believe that training the mind is at least as beneficial as training the body. We need only accept the challenge.

There are now quite a number of good books about TOC that range from Dr. Goldratt's conceptual books to a TOC handbook. This book provides a short overview of the basic concepts and then moves immediately to the virtual experience provided by the stories and analyses.

In perspective, my efforts in teaching TOC have been directed toward creating learning platforms. This is my personal interpretation of Dr. Goldratt's use of the Socratic approach. Dr. Goldratt uses the Socratic approach by raising questions and leading the listeners and readers to arriving at the solution by themselves. In this book I try to provide problematic situations: first let the reader ask the questions, then look for answers, and then compare them with the questions and answers I have provided in the book. There is a clear dominating need to raising the right questions. I remember an embarrassing moment for me while teaching a course, when one of the participants said "Eli, you raise more questions than answers." I wasn't certain how to address that remark. Was it a compliment or a criticism? You see, in my opinion, raising the right questions is far more important than the answers. Why? Because the questions are much more generic than most good answers. When one learns to ask a good question, he has gained a much better understanding of the situation. If one better understands the situation, the answer will be better. Answers are often good only in the context of the specific situation. Questions are the guidelines for our thinking. I hope this book will strengthen our capability to ask the right questions.

I have lived and thought about the Theory of Constraints for the last 13 years. Some of the more knowledgeable people on TOC may be surprised to see that I tend to take liberties in the use of it. Some of my definitions and rules deviate from what has been written and taught on TOC. I am fully aware of that. I only claim that what I have learned from Dr. Goldratt has been assimilated for some time to allow me to come up with some minor modifications that work better for me. TOC has matured enough to allow for some disagreements and a more free format use of the thinking tools.

Several people have helped me through this writing experience. First of all, I had great assistance from Bill Dettmer, himself a noted author on TOC. Dr. Alex Klarman, a partner with the Goldratt Institute, has provided very valuable feedback, as did Chanan Lechtman, an experienced Israeli implementor of TOC. Dr. Avner Passal is one of the leading Israeli organizational consultants with whom I have worked and collaborated extensively. Together we have used some of the stories as part of our workshops. Together we have thought about and developed how to link behavioral concepts such as organizational culture into the domain of TOC.

As I have been trying all those years to expand the discussions about management and dilemmas, let me mention more "independent TOC thinkers" with whom I enjoy arguing, because some real learning is generated: Prof. James Cox and Robert (Rob) Newbold. Last, but certainly not least, is Avraham Mordoch, who like myself is a former partner with the Goldratt Institute and an excellent devil's advocate to almost any new idea I have. Thank you all.

Eli Schragenheim
Ra'anana, Israel

The Author

Eli Schragenheim is an active management consultant in Israel. He joins forces with consultants from different knowledge bases such as organizational behavior and engineering. He has delivered hundreds of workshops for managers, presented several papers at academic and annual professional conferences in the U.S. and the U.K., and taught operations management at the Tel Aviv College of Business Administration. He has published several papers in academic and practitioner journals, including the *Production and Inventory Management Journal* and *The International Journal of Production Research*. He sits on the board of directors of Israel Production and Inventory Control Society.

Mr. Schragenheim is also the president of Elyakim Management Systems Ltd. (1992), which provides management education and consultation and develops large-scope simulators used as educational tools for management. In 1998, he founded MBE Simulations Ltd., of which he and Moshe Yerushalmy are joint managing directors. The purpose of MBE (Management by Experience) is to develop state-of-the-art computerized simulation for management education.

Prior to founding MBE Simulations, Mr. Schragenheim served in the Israeli army and graduated from the Hebrew University in Jerusalem where he studied mathematics and physics. And in 1980 he received his MBA from Tel-Aviv University. At that point, Mr. Schragenheim decided to pursue a career in computer programming. In 1982, he established his own software company and developed a computerized package for certified accountants.

In 1985, Mr. Schragenheim teamed up with Dr. Eli Goldratt, who at the time was the managing director of Creative Output Ltd., a software company that had developed the OPT package. The OPT software was a unique finite capacity scheduling package that carried in its logic the seeds for the Theory of Constraints (TOC). Mr. Schragenheim was hired to develop a "video game to teach the OPT concepts."

Toward the end of 1986, Dr. Goldratt created the A.Y. Goldratt Institute (AGI), and Mr. Schragenheim was among the first four people in the institute. At that time, Mr. Schragenheim began to develop what is now called the "AGI Simulators," equipped with the new tool the first AGI workshops delivered.

Mr. Schragenheim became a partner with AGI. While still developing new computerized tools for the institute, he launched new workshops and delivered the first ones. He created the workshops for the new software package, called "Disaster" at the time, and prepared a special event for the international Jonah Conference in Orlando in 1989, which was nicknamed The Orlando Game.

Mr. Schragenheim resigned from AGI toward the end of 1991 to create a new generation of computerized case studies. And he again established his own development company, which created Management Interactive Case Study Simulator.

In addition, Mr. Schragenheim has studied filmmaking. He was a freelance TV director and has directed more than 150 short and medium length documentaries for Israeli TV. He specialized in short documentaries on innovations in industry and medicine.

Mr. Schragenheim is an Israeli citizen and is married to Erela, an architect. They have three children.

About APICS

APICS, The Educational Society for Resource Management, is an international, not-for-profit organization offering a full range of programs and materials focusing on individual and organizational education, standards of excellence, and integrated resource management topics. These resources, developed under the direction of integrated resource management experts, are available at local, regional, and national levels. Since 1957, hundreds of thousands of professionals have relied on APICS as a source for educational products and services.

- **APICS Certification Programs** — APICS offers two internationally recognized certification programs, Certified in Production and Inventory Management (CPIM) and Certified in Integrated Resource Management (CIRM), known around the world as standards of professional competence in business and manufacturing.
- *APICS Educational Materials Catalog* — This catalog contains books, courseware, proceedings, reprints, training materials, and videos developed by industry experts and available to members at a discount.
- *APICS — The Performance Advantage* — This monthly, four-color magazine addresses the educational and resource management needs of manufacturing professionals.
- *APICS Business Outlook Index* — Designed to take economic analysis a step beyond current surveys, the index is a monthly manufacturing-based survey report based on confidential production, sales, and inventory data from APICS-related companies.
- **Chapters** — APICS' more than 270 chapters provide leadership, learning, and networking opportunities at the local level.

- **Educational Opportunities** — Held around the country, APICS' International Conference and Exhibition, workshops, and symposia offer you numerous opportunities to learn from your peers and management experts.
- **Employment Referral Program** — A cost-effective way to reach a targeted network of resource management professionals, this program pairs qualified job candidates with interested companies.
- **SIGs** — These member groups develop specialized educational programs and resources for seven specific industry and interest areas.
- **Web Site** — The APICS web site at http://www.apics.org enables you to explore the wide range of information available on APICS' membership, certification, and educational offerings.
- **Member Services** — Members enjoy a dedicated inquiry service, insurance, a retirement plan, and more.

For more information on APICS programs, services, or membership, call APICS Customer Service at (800) 444-2742 or (703) 237-8344 or visit http://www.apics.org on the World Wide Web.

A word about case studies is in order here. The Harvard Business School publishes perhaps the most complete listing of business case studies anywhere. Most of those case studies describe things that actually happened to the organizations written about. Many of them use the real names of these companies; some are disguised to protect the guilty. As you might imagine, not all the cases paint the organizations described in a favorable light—but as one philosopher once said, "None of us is completely useless; even the worst of us can serve as a bad example!" Either way, however, the impact of the lesson—the principles and concepts under discussion—is considerably more powerful with a real world example to demonstrate it.

Every Harvard Business School case study has a standard disclaimer printed on it. This disclaimer says, in part, "[this case was prepared] as the basis for class discussion, rather than to illustrate effective or ineffective handling of an administrative situation." To which I say, "Horsefeathers!" These cases are like Aesop's fables: If there is not an underlying lesson to be learned, then why bother discussing them? That disclaimer, clearly intended to impose some indemnity from legal action, if taken seriously will preclude any significant learning and make intellectual discussion about it a waste of time.

You will not see this in any of Eli Schragenheim's cases. These stories are definitely intended to convey specific lessons about the application of constraint theory. Eli has collected a wide range of cases, encompassing all kinds of organizations—for-profit, not-for-profit, commercial, government agency, and so forth—expressly to demonstrate the applicability of constraint theory and its tools to any kind of complex system. The discrete nature of the application might change from one situation to another, but the applicability remains universal.

So read this book with the following guidelines in mind. First, look for the application of constraint theory in each case. Eli thoughtfully allows you to do this on your own. But for those who have difficulty seeing the application, he then provides his personal analysis of the situation and an explanation of how constraint management principles fit in. A word of caution is in order, however. Although Eli's first chapter contains a succinct explanation of constraint theory, its concepts and principles, this book offers far more educational value if the reader has already read the books of Goldratt and others. Do not let that stop you from proceeding. On the contrary, reading this book before the others available on constraint theory can bring those other books much more to life.

Second, and finally, we all like to believe that we are unique, different. And to a great extent, we are. But despite our differences, circumstances and organizations have much about them that are similar, even if they seem different on the surface. Yet this desire to be distinctive often motivates us to say, "Well, that's great, but we're a *service* organization, not a *manufacturing* one; how could that possibly apply to us?" You can substitute any other pairs of words for the ones in italics; the song is the same, even if the lyrics vary. I would challenge you to read this book with a different mind-set: See how many similarities you can find between these stories and your situation, not how many differences. Being different is desirable if you are trying to differentiate your product or service from your competitor's. But when you are trying to make your system work better, it is often a lot easier to see how previously developed solutions might be tailored to your circumstances than it would be to build a new one from scratch. So read on—and enjoy! This book is one you cannot afford to miss.

H. William Dettmer
Author, *Goldratt's Theory of Constraints: A Systems Approach to Continuous Improvement* and *Breaking the Constraints to World-Class Performance*

1 Basics of TOC and How To Analyze a Variety of Cases

The Theory of Constraints (TOC) is a unique management philosophy that strives for a rationale or scientific approach to management. It provides a way to simplify the complexity of human-based systems and still keep the main issues and impacts under managerial control.

In our voyage to understand the TOC way of management thinking, let us start with a short story. We will use this story as a platform to demonstrate the basic TOC ideas, tools, and my own approach to the analysis of a case. As TOC was first developed in manufacturing, I have intentionally chosen a nonmanufacturing story to begin with, to demonstrate at the outset that TOC is not restricted to manufacturing applications alone.

The story shows us a failure. Failures provide us with an opportunity to learn. The first thing that either the manager in charge or the consultant who was called to help should do when confronted with an unfavorable situation is to understand the cause and effect that led to the problematic situation. If you do not do that, you may come up with a great solution to a problem that does not exist. In other words, the unfavorable situation will persist. Once such an understanding is established, the next challenge is to find a good solution. So, in this story, as well as in the rest of the stories in this book, the first mission is to identify the main causes and only then look for solutions.

My brief personal overview of TOC follows the story. Then there is a description of how I would approach the situation using TOC tools. Please be aware that I offer no guarantee that I am right; you may disagree with me on the analysis and the proposed solution. As long as you can back up your analysis of the facts in the story using your intuition, life experience, and common sense rationale, that is great. There can be more than one way to solve a problem. TOC does not provide "canned" solutions. Rather, it provides the methodology to develop your own effective solutions.

So, here is the first case, a relatively large example.

A Hotel That Does Not Fit In

Roy has planned his new business initiative very carefully. He decided to move into the tourism and resort business, starting with his own hotel at Lake Yuma. Beautiful scenery, peaceful weather, easy access, and old-fashioned, expensive hotels are the only competition. Roy thought that establishing prices 25 percent lower than his competitors would attract many more visitors throughout the year. Coupled with his ideas for keeping costs down, he planned an impressive facility: 400 rooms, two convention halls, a beautiful private beach, and a gourmet restaurant, which was considered an important key to success.

The G-Roy Hotel opened in March 1997. The advertising campaign worked very well. Rooms for the first six months were almost booked up even before the grand opening. Another financial success seemed to await Roy. However, demand dropped sharply. By September, occupancy had dropped to a mere 50 percent compared with the 70 percent occupancy of his conservative competitors during the same month. Because Roy's whole business strategy was to fill up the hotel during the off-peak periods, the prospects of the G-Roy Hotel seemed gloomy indeed.

Roy sat in his office alone. All the reports about the G-Roy Hotel's activities were laid out before his eyes. He had no doubt that the answer to the question "What really went wrong with G-Roy?" was buried somewhere in the huge file of customers' complaints rather than in the business analysis his assistants wrote. The complaints were arranged in categories. There were 204 complaints about the restaurant, 166 complaints about the slow response of the elevators, 127 complaints about the small size of the rooms, 125 complaints about the queue at check-in and check-out, 94 complaints about the cleaning and service of the rooms, 10 complaints about malfunctions of some devices in the rooms, and 7 complaints about improper language by certain employees.

Roy thought about the basic assumptions behind his planning. He assumed that the customers of such a nice resort would be willing to have smaller rooms than customary in exchange for a substantial reduction in price. He was quite prepared for the complaints about the room sizes. It was the other complaints that surprised him. The biggest surprise was the restaurant. The vast majority of the complaints were that it was very difficult or even impossible to get a table. Roy had visited many resort hotels and found that on average, only 53 percent of the guests used the restaurant for dinner or lunch, during the two hours when most people eat their meals. He decided to plan the restaurant for 58 percent of the total occupancy of the hotel. Still, people complained that they had to wait too long for tables, even though a careful check had confirmed that only 53 percent of the guests tried to eat at the restaurant during the peak dinner times.

Roy felt angry. He knew the clients were right, but he had tried his best to establish an excellent dining facility. He had hired two notable chefs to run the restaurant. He paid them enormous salaries, but all he got in return were complaints that the kitchen was too small and the pressure was too high.

The elevators were a similar story. The three elevators were carefully calculated to serve 400 rooms on 12 floors. Why were they not enough? Sure, in the first three months there were some troubles with one of the elevators, but that elevator was fixed. The number of complaints was somewhat reduced, but new complaints continued.

So, what is wrong with the G-Roy Hotel? Is it impossible to design a hotel that will not simply waste money?

The Classical TOC Basic Concepts

Let us set the story aside for awhile and search for some broad guidelines for management. Is there a universal problem that all managers face?

There are two cardinal universal problems that are interrelated. The first is that any organization that has more than just a few people is, above all, a complex system. The term complex means that it is difficult to predict the outcomes of a given action across the organization. The causes and effects are not easily identified simply because of the overwhelming number of interactions among people and departments. The second problem is the uncertainty around and within the organization. That means that any decisions or actions we choose may produce different results.

Those two causes produce the effect that it is difficult to know what the best decision is in different situations. That is what makes the role of the manager more of an art than a profession that can be learned and mastered through formal studies.

TOC offers a way to deal with the issue by creating a picture of the organization that is "good enough" rather than precise, one that will simplify the complexity. Once a simplified version of the organization exists, the uncertainty can be dealt with by implementing appropriate protection mechanisms at the most critical areas.

As part of its scientific approach, TOC is based on certain assumptions and logically develops a set of guidelines that are logically derived from that set of assumptions. In my interpretation of TOC, the following three basic concepts are key assumptions in understanding the TOC philosophy:

1. An organization has a goal to achieve.
2. An organization is more than the sum of its parts.
3. The performance of an organization is constrained by very few variables.

Let us look into each of those leading assumptions. The goal of an organization can be described as the single objective that an organization wishes to increase or improve. In achieving its goal, an organization must meet a set of conditions that vary from one organization or industry to another. Although those necessary conditions must be fulfilled to achieve the goal, they themselves are not the ultimate goal.

For example, say a business organization wishes to make as high a profit as possible. It may establish rules of business behavior, such not allowing bribery even when dealing in countries where bribery is acceptable standard behavior. The organization's goal is still to make more money but under the restrictions of certain business ethics. Another necessary condition might be that the organization requires a certain level of commitment from its employees and in return must provide a certain level of satisfaction. That does not mean the organization wants to maximize the employees' satisfaction but that it must provide a minimum satisfaction level if it is to elicit the commitment it needs. The idea of maximization can be realistically applied to only one entity, while all the other necessary conditions are characterized by a certain minimum level that the organization has to satisfy.

To understand the full meaning of the second assumption—the organization is more than the sum of its parts—we should realize that the ability of an organization to achieve a common goal depends on the synchronization of its parts in a combined effort. That means that we cannot divide an organization rooms. However, the total revenue is the product of the number of rooms and the rate per room. That rate is determined by the market demand.

Also, the goal is to make money now as well as in the future. The number of rooms limits the current profits. The market demand limits future sales, assuming it is possible that not all the rooms are taken in the future.

Why was it so difficult to subordinate to the guests' requirements? Is it because the personnel were not professional enough or lack that special attitude about service? Roy's careful planning indicated that he was mindful of hiring very professional and experienced employees. If he planned a high quality restaurant and hired highly qualified chefs, cannot we assume he also demanded a high level of service from all other employees? That was not where Roy tried to cut costs.

The probable answer is that Roy planned too little excess capacity on all resources. Let us consider the restaurant. The decision about allocating space to the restaurant was crucial. Roy relied on statistics about average use of restaurants across the industry. He padded that figure only slightly. However, averages are only that. They imply that sometimes the demand for restaurant seating will be much higher and sometimes much lower. At any given time, many more guests may be lining up for meals, and excessive waiting may seem to be bad service. Of course, if that happens once a year, it is not so terrible. But when it happens three times a week, it is another situation altogether.

The same thing pertains to the front desk. There should be enough clerks to allow for an appropriate response—meaning a minimum acceptable waiting time—at any time. That means the peak demand should dictate the number of clerks. Planning based on average demand always equates to lousy subordination to the guests.

The same applies to the elevators and the cleaning of rooms. The state of the elevators may be devastating because it may add to the pressure at the restaurant. When the elevators cannot respond quickly to all the people who want them, queues build up. Every elevator travels full, so several successive full elevators all delivering guests in large batches can quickly overwhelm the capacity of the restaurant to seat them, and long lines form at the maitre d' podium.

The evils of lousy subordination negatively affected the future market of the G-Roy Hotel. How did such a shrewd businessperson fall into this trap? After all, anyone who knows even a little about statistics ought to realize that the service level should be calculated for the average, plus a certain number of standard deviations. How many standard deviations should be added to the average needs to be based on the damage done to customer expectations by failing to provide excellent service.

Statistics is not a very robust decision support system when the data is not available. Roy could, perhaps, have roughly estimated the standard deviation for the use of the restaurant during peak times. But that still would not tell him how long customers are willing to wait before they decide never to come to that hotel again. That input is critical for calculating the optimum size of the restaurant. When it comes to the queue at the front desk and at the elevators, it may be even more difficult to come up with the "optimal" decision, since the data that describes the distribution for such queues is more difficult to obtain.

We can understand Roy's desire to calculate the optimal size of the restaurant, the number of elevators, and how to run the front desk in the most efficient way. On one hand any excess space at the restaurant, excess elevators, and too many front desk personnel can be perceived as costing quite a lot of money, making it more difficult to produce a profit. On the other hand, having too little space, too few elevators, and not enough clerks to handle a busy front desk can limit future sales and profits.

This is a basic conflict in management: How to keep costs down while making current and future sales high as they can be. No true optimization is possible in an environment in which not enough data is recorded, the recorded data is not accurate, and a large number of parameters are involved. If we had all that information, we would still need a mathematical miracle to come up with the optimal solution. In reality, we do not have all that data or the mathematical super-formula, and we have a lot of uncertainty that adds even more to the hopeless search for optimization. In such an environment, there is not much point in making an effort to record more and more data. We need to find another way to come up with good decision making support.

TOC offers a way to get out of the data collection trap. It permits us to use merely "good enough" information yet still translate results all the way to the bottom line. To achieve positive bottom-line results with imperfect information, only a very few constraints must be maximized while all the other elements of the system must have excess capacity and capability. In the case of the G-Roy Hotel, that means the restaurant, which should not be the hotel's constraint, must have somewhat more space than seems necessary even for a peak season. Certainly there should be one more elevator than what originally seemed to be enough. The same reasoning applies to registration and check-out.

Subordinating properly requires enough excess capacity and capability in all other parts of the system so that performance at the constraint is not threatened. In the G-Roy Hotel, all the supporting elements—elevators, restaurant, and front desk—must have enough capacity so that quality of service is never compromised, even if all the hotel's rooms, the ultimate resource constraint, are fully occupied.

That might mean investing more money than initially thought in order to preserve and improve revenue. In TOC, we refer to this concept of emphasizing the increase of revenue over the reduction of costs as operating in the "Throughput World." It is one of the most important prescriptions of TOC. To truly understand the impact of the Throughput World, we must address the role of global measurements in the Theory of Constraints.

The Throughput World Vs. the Cost World

As we saw earlier, the first assumption of TOC involves the existence of a goal. How good are we in achieving that goal? The answer to that question is an important input to our decision making. This is what measurements enable us to do: Judge the consequences of actions taken and guide decisions about what actions to take.

Why not use the definition of the company's goal as our primary measurement? That would mean that every decision and action will be measured according to its impact on that global goal.

For example, suppose, a tour operator suggests that the G-Roy hotel reserve 10 rooms throughout the whole year for a mere 50 percent of the regular rate. How should the hotel manager go about deciding whether or not to do this? Does the net profit measurement provide a clear choice? Does return on investment (ROI)? The indirect way net profit and ROI are calculated makes it very difficult to see which choice is unequivocally preferable. What effect, for example, will the filling of those 10 rooms at half the normal rate have on long-term net profit? The connection is anything but clear.

TOC provides a set of three measures that are better suited to dealing with daily decisions and better match our intuition. Those three measures are called throughput (T), inventory (I), and operating expenses (OE).

Throughput is the foremost measurement in TOC. It does have some parallels in the other management theories, but none has put so much weight and thought into this term. Assuming an organization's goal is to make more money, throughput is defined as the rate of money generated by an organization.

The organization makes money by selling products or services. Throughput is looking at the single sale. Every sale generates revenue and causes some expenses. The throughput generated by a single sale, thus contributing to the total throughput in a period, is:

The revenues that were made from the single sale minus the truly variable expenses that are caused by that sale.

What is the idea? When the organization sells something, it should add value to the goal. That added value is the revenue from that sale minus those expenses that were caused by that sale. That means if we were not selling this particular piece, that added value would not have materialized.

What is the throughput from a guest paying $100 for a room? In this example, we may conclude that the truly variable expenses are practically zero. If you wish to be very precise about it, you may claim that the guest soap and shampoo, the soap used for the laundry of the linen, and the cost of electricity and energy are all incurred by renting the room to a guest for one night. Assume all these truly variable expenses of a room comes to $1 per night it is actually used. That means the throughput of renting that room for one night is $99.

Of course there are many more expenses that need to be considered. For instance, the room must be cleaned. The hotel pays the maids that clean the rooms. The guest occupies some of the time of the people at the front desk. Should we include some portion of the personnel salary to get a more realistic throughput for a room per night?

TOC suggests that we should not. Those expenses are not directly related to the throughput generated by having a customer in the room—their relationship to throughput is only indirect. The personnel salaries will not change if that customer had chosen not to come. The indirect costs do have a bearing on overall net profit, so they cannot be ignored completely. But TOC segregates fixed expenses from variable ones and accommodates them later in the profit calculation.

This treatment of throughput, taking into account only the costs that are directly caused by a single sale, is what makes the TOC approach to accounting so beneficial. In economics, the term "contribution" is used to define something basically identical to throughput. However, contribution usually subtracts the direct labor cost from the revenue. Throughput's definition, according to TOC, does not include direct labor unless the direct labor really varies with the sale of every unit. The importance of the concept of through-put lies in its ability to support decisions by predicting how much those decisions add to the bottom line. We shall return to this concept and see its impact on decision making.

Besides throughput, TOC establishes two additional measurements to complete the global set:

Inventory or investment (I): All the money the system invests in purchas-ing things the system intends to sell.

Operating Expenses (OE): All the money the system spends in turning inventory into throughput.

To generate throughput, we need to invest money in the system. This capital that is held in the system is called inventory. The traditional definition of inventory is interpreted here as an investment. But unlike generally accepted accounting procedures, works in process and finished goods are measured by their original acquisition value only, without any added value representing the labor that has thus far been invested. TOC suggests that value is truly added only when throughput is generated, meaning when the expected revenues materialize. At that time, the cost of purchasing the materials is deducted from the revenues as part of the truly variable expenses per sale.

To generate throughput, we need to pay operating expenses. All costs that are not directly incurred by a sale constitute operating expenses. Generally, they are fixed costs, such as overhead. Unlike traditional management cost accounting, TOC considers labor a fixed cost, too, because salaries are paid whether or not the product is sold. From those definitions we can derive a relationship between throughput, operating expenses, and net profit:

$$\text{Net Profit (NP)} = \Sigma T - \Sigma OE$$

Now you may ask: If this is that simple, why is it more useful than any other set of measurements? First, any decision made just on a single sale impacts only the throughput element of the equation, not the operating expenses. That makes it easier to evaluate possible decisions. It divides the business into two sensible parts, the added value generated and the pool of expenses that is needed to generate that added value. Though most TOC practitioners separate inventory/investment and operating expenses, I prefer to view the inventory/investment measurement, with the support of financial models, as a stream of operating expenses. That focuses every decision on the change in the throughput relative to the change in the operating expenses. In other words, you consider the benefit (throughput) vs. the cost.

Assume that the hotel manager considers the option of selling 10 rooms for a whole year for half of the usual price to a certain travel agent who is fully committed to filling those rooms. The decision does not impact either the inventory or the operating expenses of the hotel. There are two possible effects on the throughput. First, selling the 10 rooms for the whole year generates additional amount of throughput. If the suggested price is $50 per night and there is $1 worth of materials, the total generated throughput for the whole year on 10 additional rooms is: ($50 − $1) × 10 × 365 = $178,850.

But because the hotel is constrained by the number of rooms at least during some periods of peak demand such as holidays, then the hotel loses the throughput of those 10 rooms that could have been sold to other customers at full price. Suppose the hotel management evaluates that throughout the year, there are 100 nights that all the available rooms could be rented for the regular price of $100. By making the deal with the travel agent, throughput is lost for: −99 × 10 × 100, because those 10 rooms have been already sold. The net change in throughput is $79,850. This analysis makes it clear that the offer is a good one.

Of course, if the hotel management can achieve full occupancy on its own throughout the year for the regular price, then accepting the deal will reduce the total throughput and is not desirable.

Suppose instead the offer was to fill 100 rooms at the reduced price. How does that change the analysis? Assuming extensive off-peak periods do exist, the management should ask itself whether maintaining additional 100 rooms during off-peak season would add operating expenses. Because some hotels use a temporary workforce, that may be relevant to the decision. Another question is whether the 100 rooms are available during the off-peak season or are some of them rented at the expense of other better deals or full-paid customers. In short, with such a deal, the number of rooms may be an active constraint for longer periods of time. The generic conclusion we can draw from this example is that a relatively big decision may have a direct and immediate effect on the identity of the constraint.

The notion of throughput raises the following question: Where should the main efforts of management be directed? To increase throughput, reduce inventory, or reduce costs?

Is it a strange question? It seems natural to expect top management to look at the three avenues that all improve the bottom line: Throughput, inventory, and operating expenses. However, in reality improvement efforts are constrained by the management's ability to lead many changes. At the end there is a clear need to **focus** on the decisions and actions that will most quickly show an improvement in the bottom line.

Consequently, most managers first try to reduce costs. Even the total quality management (TQM) and the business process re-engineering (BPR) movements sell their methods based on their capability to reduce costs. Most managers consider reducing cost to be more under their control than increasing throughput or reducing inventory. Moreover, it is easier to set an exact objective in cost reduction (for instance, reduce costs by 6.9 percent). So the preferred way for most managers to improve profitability is to reduce costs. That was Roy's focus in planning his new adventure—keep costs down. It did not work in this case. It does not work in too many others either.

There are two problems with focusing on cost reduction. One is that in keeping management busy "right-sizing," the options for a dramatic increase in throughput are often overlooked. The second problem is even more devastating. In the rush to reduce costs, it is easy to disrupt the delicate line between "excess capacity," which can seem like a waste of money, and "protective capacity"—the level of unused capacity that is needed to permit appropriate subordination to the market and/or to the internal constraint.

That is what happened in the planning of the G-Roy Hotel. Although there was some unused capacity in the restaurant, it was not enough to ensure subordination to the full number of guests that could be accommodated in the rooms. On paper, the number of elevators seemed just right. In reality, "just right" can easily turn into "definitely not enough."

The Throughput World is a name given by Dr. Goldratt to the part of TOC thinking that recognizes the goal of an organization and strives to get as much throughput as it can squeeze out of the constraint. That is very different from the Cost World mentality, which emphasizes cost control above all else. All the cases in this book were chosen to train the reader in Throughput World thinking.

Effective application of TOC requires us to emphasize increasing throughput as a first priority. The rationale behind the Throughput World is that theoretically, there is no limit to the throughput that can be generated, as constraints can be elevated. And every time that is done, the organization stabilizes at a higher level of throughput and higher level of profits. On the other hand, we can reduce the operating expenses to zero—and that is it. Of course, with zero operating expenses the throughput will be zero as well.

TOC and Cost Accounting

From the three basic assumptions, a challenge to some of the most sacred truths of management emerges. No matter what kind of cost accounting you consider, the term product cost is a basic term. It may appear easy to calculate the costs of a product unit, considering all costs generated in producing one unit gives us a great decision support. For instance, in directing the sales-people whether to give price reductions, the decision rule is easy: Don't sell at less than the cost. The product cost is also a critical figure in specifying the price for a bid or promotion and sets the "cost plus" relationships with contractors. When we know the costs, we know when we are making money or losing money.

The Question is, Do We Really Know the Cost of a Unit of a Product?

The product cost term can be useful only if it is independent of all the other products of the organization. Otherwise, for any decision we must consider the change in all the product unit prices due to that decision. In such a case, we had better consider the cost of the whole quantity. Why? Is the cost of a product unit independent of the other products?

Since some resources, such as marketing, accounting, and general man-agement, will be common to several products, the answer is no! Only a holding company of several unrelated manufacturing companies could have products of one company that are *really* independent of the products of other companies. In most cases, there is a partial dependency between the various products produced by the same manufacturing company. This is what the second basic assumption is all about: The organization contains many inter-dependencies.

Let's go back to the hotel and look at the cost of room service. Any meal delivered is a product in itself. Traditionally, the cost of such a delivery includes the material cost and a whole set of allocated cost activities: accepting the order, cooking and preparation, delivery, taking the tray back, and cleaning the dishes. One can calculate the cost of the labor involved with these operations.

In addition, there are many other indirect expenses to be tied to this product: the overhead of keeping the kitchen and bar, the purchase of food, the management of the staff and the management of internal communication. What about the infrastructure of the hotel? What part of that is to be allocated to a single delivery of room service? There are many ways to allocate indirect cost to a single delivery. Any traditional cost allocation, as well as activity-based costing, will include part of the costs generated in the kitchen as part of the costs allocation to any room service delivery.

The kitchen represents a significant overhead expense in the operation of the hotel. But the kitchen should always have enough capacity to handle all the demands of the restaurant guests. Certainly it would look real bad if when a queue of people is lining up, the food service is slow as well. If we assume that the kitchen has some excess capacity that helps to respond adequately to the restaurant guests, then the additional meals for room service can be prepared without incurring extra cost except for the food. Continuing this line of reasoning, we realize that additional room service deliveries generate throughput without any change in the operating expenses (the food is a truly variable cost and is deducted from the revenue as part of the throughput calculation). As a matter of fact, we can see that part of the solution to the lousy subordination problem of the G-Roy Hotel might be to promote room service to reduce the demand for tables in the restaurant.

If any of the kitchen costs are assigned to a room service delivery, it will guide the management to price the service accordingly. In the situation stated above in which there is intense pressure on the restaurant tables, we might find that the room service meals are priced well above the restaurant prices, causing people to come down and line up at the restaurant. Instead of promoting the room service—a policy that would improve the profitability— the product cost approach drives management to make the **wrong** decision: Pricing room service in a way that discourages its profitable use while simultaneously swamping the restaurant with more customers than it can handle.

The tendency to calculate a phantom called "product unit cost" is one of the major policy constraints that is so common in the business community. A policy constraint is first of all a constraint—something that limits the performance of an organization. Then, because it is also a policy— a procedure created by the management—it is a constraint we formally impose upon ourselves, usually without realizing it!

Why Should Any Management Create a Policy That Limits the Performance of an Organization?

There are two common causes. One is inertia. That means the policy might have been beneficial at the time it was created. However, as time has advanced and the environment (and probably the organization) has changed, the policy is no longer useful. It may in fact be constraining the whole organization. The second cause is that the policy has been derived by a thinking paradigm that is based on an erroneous perception of being in a conflict between two requirements. TOC provides a set of tools called the thinking processes to analyze the current situation, identify such policy constraints, and devise the route to significant improvement.

TOC Thinking Processes (TP)

In this overview of TOC, I do not intend to explain all five formal thinking processes and the procedures to use them. I will concentrate primarily on the three tools that will be handy in the analysis of the stories. Even those tools are given here a less pedantic treatment than is usually applied. For a much more rigorous approach, please refer to other books such as H. William Dettmer's *Goldratt's Theory of Constraints: A Systems Approach to Continuous Improvement.*

The G-Roy example was analyzed using intuitive logic guided by the three concepts and the five focusing steps. That intuitive logic is an asset we cannot live without. However, we can strengthen it by verbalizing our arguments, assumptions, and hypotheses. When we further put the arguments on paper in a form that is easy to follow, we open ourselves to two different sources of enhancements. First, we can read and check our own logic again. Second, we can present it to others who can look at every facet of our logic and raise their reservations about it.

The reluctance of so many people to present their logic clearly and openly stems from the danger of exposure. We often protect ourselves in a cloak of ambiguity. When one's arguments are ambiguous or too generic, it is difficult to put a finger on any flaw. All that is left is to say, "I disagree." By putting our intuitive logic on paper, we become exposed. The benefits can be enormous. The threats to one's ego are still real, but they can be reduced because once it is possible to train our own thinking, our thinking will be improved. These basic ideas about logical thinking are expressed in the graphical form of the TOC thinking process (Figure 1.1).

The format of this logical mapping of the advantages and disadvantages matches the future-reality tree (FRT), one of the five tools of the thinking processes. Every arrow connects two statements. The argument is that the statement at the bottom of an arrow causes the statement at the arrow head. When the coexistence of two or more causes are needed to produce an

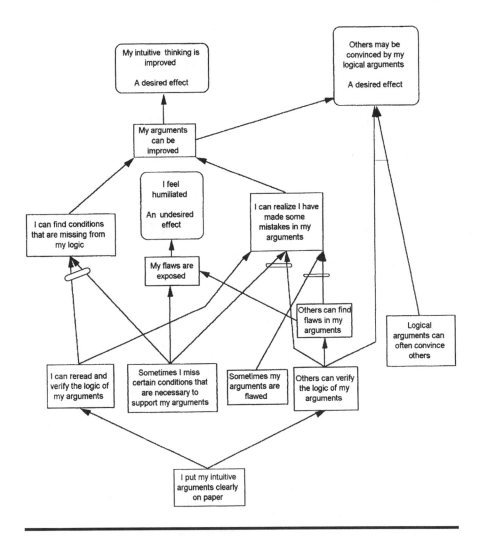

Figure 1.1   Mapping the Intuitive Logic on Paper

effect, then the several arrows are connected by an oval shape. Remember, as initially presented, this is merely an argument! It may be wrong. It cannot be considered correct until the causal relationship is validated. Until then, it is simply a mapping of various arguments into a structure that can be viewed in two dimensions, rather the single dimension of written text.

The future-reality tree is a tool to analyze the comprehensive outcomes of an idea, action, or even an external event that may happen. Another tool, the current-reality tree (CRT), depicts the current state of an organization, with the objective of identifying a root cause for the lack of success of the organization in achieving its goal. Let us go back to the G-Roy problems as seen by Roy, apply the arguments we have used to identify the problems, and see whether we can use the thinking process to develop something more suitable for analysis of the current situation.

The formal TOC/TP procedure for drawing the current-reality tree starts with a list of undesirable effects. A network of cause and effects, including other observed effects, is then built backward until all the undesirable effects are connected to one or a very few root causes.

I prefer a somewhat different approach. The most obvious undesirable effect is a lack of success, sometimes even failure, in achieving the goal of the organization under discussion. This is where I prefer to start. Thus, the identification of the constraint, physical or policy, is the intermediate objective. But that is not enough. Although every system has a constraint, one should always ask whether this is the constraint we would prefer to have. Even when the answer is positive, the question of whether the constraint has been effectively exploited must be answered.

The main undesirable effect at the G-Roy Hotel is the realization that the financial prospects are gloomy. Why? We can attribute that directly to the decrease in market demand for G-Roy's rooms. So, for the G-Roy Hotel, the constraint lies in the market. That is certainly true in September. Is this a desirable condition? Certainly not. Hence, we need to understand what caused the drop in customers at the hotel.

Where do we stop when building such a tree? The formal TOC/TP approach is to reach the level at which all the undesirable effects are connected to very few root causes (statements without a preceding cause). And if we eliminate one of those, it will eliminate the majority of the original undesirable effects. That is the core problem.

My objective is to reach a core problem that is conceptual. Hence, the current-reality tree in Figure 1.2 does not stop at the failure to subordinate but continues to identify deeper causes until the dominant paradigm—a thinking pattern—is identified.

Figure 1.2 displays my beliefs about the G-Roy situation. The reader should verify the logic to see whether it makes sense. Another important check is to determine whether this method of analysis and communication of logic is better than traditional methods of constructing documents. The same should be done for all other analyses of stories in this book where this representation is used.

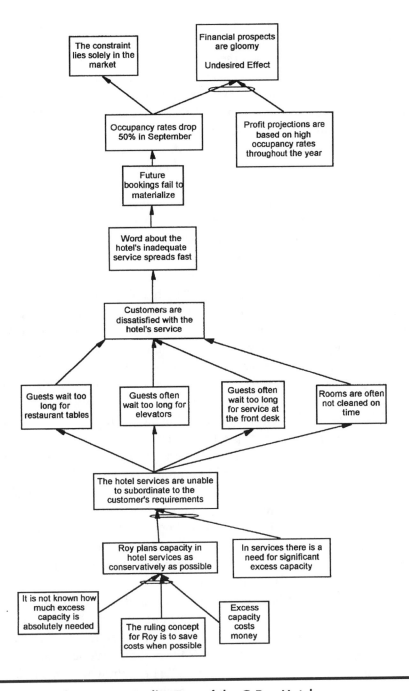

Figure 1.2   The Current-Reality Tree of the G-Roy Hotel

This current-reality tree proposes that Roy's concern for keeping costs as low as possible may be the basic cause of the gloomy financial prospects of the hotel. However, there is another core cause. Certainly Roy could not have known, at least at the planning stage, *exactly* how much excess capacity is needed to ensure an adequate service level. If there were a mathematical/statistical formula that could calculate exactly how much excess capacity should be planned for any enterprise, then it would be much easier to be a manager, and even easier to learn to be one. But the exact shape of the uncertainty (the distribution function) is usually not known. The dependencies among the various parameters are difficult to model, and too much work is needed just to collect the data. Hence, real-world managers must be satisfied with approximating the amount of excess capacity needed to account for uncertainty. In this case, Roy tried to save costs as much as possible and underestimated the need for excess capacity.

A question that is not answered in the current-reality tree is: Why is it so important to save costs? That seems to be a trivial question—obviously, we need to save costs to make a profit. But there is a "down" side to cost reduction, too. If we cut costs too much, our ability to subordinate may be less than adequate. So there seems to be an inherent conflict in the planning stage: Plan for low costs or plan for good subordination.

Whenever our analysis identifies a core problem, there is a high probability that the problem can be verbalized as a conflict between two forces, each driving us in opposite directions. If this were not the case, then why is it we cannot simply do the opposite of what the core problem states? The only case in which a problem cannot be verbalized as a conflict is when that specific problem was not identified as a problem at all. Suppose a business is constrained because it has only one telephone line, which is busy all the time. The only reason for not ordering more phone lines is because somehow no one has recognized this as the underlying cause of the business's limited performance. In a situation such as this, once the phone line shortage is recognized, the solution—adding more phone lines—is easy and immediate because it is not even perceived as a change.

A certain action might produce benefits but at the same time also lead to damage. That constitutes a conflict, and the most common response is usually to find an acceptable compromise. This way, the conflict stays on as a problem that is tough to crack. Sometimes that is the best that can be done. In many cases, though, it is possible to find a *much* better solution—something that keeps the benefits and eliminates the damage.

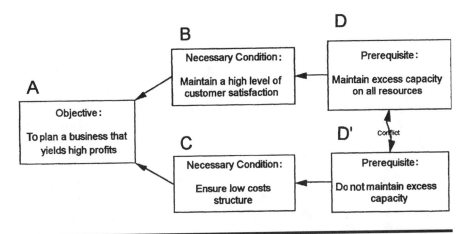

Figure 1.3   The Conflict Resolution Diagram: Costs and the Need for Excess Capacity

The idea of saving as much cost as possible also can be represented as a conflict. Analyzing conflicts can lead to surprising solutions that are not based on compromises between the two conflicting forces, but rather on an idea that both sides can actually "win" in the conflict resolution.

Dr. Goldratt developed a tool for verbalizing conflict and creating superior solutions: the evaporating cloud (EC). Although this name emphasizes the idea of finding ideas to evaporate the cloud, I prefer the name given to it by Dettmer: conflict resolution diagram. Throughout the book I will use both names interchangeably.

Figure 1.3 is a graphical representation of Roy's conflict involving saving costs. It is an example of the generic internal conflict between the Throughput World and the Cost World regarding long-term strategic planning.

First, a word of caution. The role of the arrows in Figure 1.3 is different than in the other logic trees. When the future-reality and current-reality trees were introduced, the notion of A→B meant if (entity A exists) then (it causes entity B). In mathematics, that means that A is a sufficient condition for B to be caused.

But, in this tool, B→A means that to achieve A, we must have B. B may not be sufficient to cause A alone, but if there is no B, then there is no A. In mathematical terminology, B is a necessary condition for A.

In Figure 1.3, A is the **objective**. In this example, the objective is a future business that will generate high profit. To have such a business, we must maintain a high level of customer satisfaction. To maintain a high level of customer satisfaction, we must allow for excess capacity of the vast majority of the internal resources.

On the other (bottom) hand, to have high profits, we must ensure low costs. To ensure low costs, we must not allow for excess capacity.

Any problem that persists for an extended period of time indicates the possible presence of a conflict such as this one. The classical solution to such a conflict is to find the middle ground—to trade off some of each prerequisite to a point that we can live with. We have traditionally come to believe that this is truly optimal, that it gives us the best results under the circumstances.

Do trade-offs work? Suppose we could know how much excess capacity we need to maintain a certain service level. For instance, 99.9 percent of the customers will not wait more than 10 minutes for a table, elevator, or front desk clerk. Does that solve the problem? Once we are sure this is exactly the amount we need, that is what we should plan for, no matter how much it costs. Right? Not quite. How do we know that waiting 10 minutes is acceptable? Is that acceptable enough for good business? Would you recommend that your friends go to a resort where they might have to wait up to 10 minutes for every service? On the other hand, maybe 85 percent of the customers would wait up to 10 minutes. And in this case we might be able to cut the total expenses by 4.7 percent. Isn't that a worthwhile trade-off? Just how much can we inconvenience the customer in our effort to save costs before we lose that customer? The problem with trade-offs, especially in customer services, is that we can never be sure exactly where the acceptable trade-off point lies.

So, although the conflict is real, all we usually do is to look for a good compromise. Unless, we can break at least one hidden assumption behind the arrows of the diagram. The preceding statement needs further explanation. Behind every arrow in Figure 1.3 there are several assumptions we have made consciously or unconsciously— but not stated. Those assumptions justify our claim that the entity at the tail of the arrow is really necessary for achieving the entity at the arrow head. If we can invalidate even one assumption, then it should be possible to achieve the condition at the head without taking the action at the base. Eliminating the need for one of the prerequisites (D or D') means that the conflict will not exist anymore. In Dr. Goldratt's terminology, it is called evaporating the cloud. It also constitutes a "win" for each side because both conditions necessary to achieve the objective (B and C) are satisfied. As a matter of fact, eliminating any arrow will evaporate the cloud.

Consider the AB arrow in Figure 1.3. Why do we believe we need customer satisfaction to be profitable? To reveal the hidden invalid assumptions, we should look for the circumstances in which there is no need for customer satisfaction to be profitable.

Assumptions behind the AB arrow:

AB1:   Unsatisfied customers have a choice. (If customers need the product and there is no other choice, they'll buy even when highly unsatisfied.)

AB2:   The organization makes money from its customers. (If the organization loses money on the sale, customer satisfaction is not desired.)

AB3:   The decision to buy is in the hands of the customers. (An environment in which this assumption is not valid: A professor tells the class to use a certain text book that is not to the liking of the students.)

AB4:   New customers are aware of how satisfied previous customers have been. (If information doesn't spread, then certain organizations may do well even though none of their customers is happy.)

You might be able to think of even more hidden assumptions.

Assumptions behind the BD arrow:

BD1:   Customers demand timely, quality products.

BD2:   Demand is very uncertain.

BD3:   Capacity cannot be adjusted immediately.

BD4:   We cannot say "no" to a potential customer (otherwise, we could refuse to sell when we are unable to meet all the requirements) or it will cause real harm to us.

Assumptions behind the AC arrow:

AC1:   Our company is not paid according to its costs. (Certain cost-plus agreements eliminate any motivation to save costs.)

AC2:   It is easy to be seduced into wasting money. (One may pay for things that are not needed.)

AC3:   The additional costs will not lead to higher throughput.

Assumptions behind the CD' arrow:

CD'1: Adding capacity generates additional costs.

CD'2: Unused capacity does not generate more throughput.

Assumptions behind the DD' arrow:

DD'1: Both sides of the conflict relate to the same capacity level for the same resources.

DD'2: There is no formula to calculate the precise optimum.

Any assumption is a target for a new solution that may "evaporate" the cloud. The basic prescription of the Throughput World is to challenge both AC3 and CD'2, especially CD'2. If throughput can be elevated by effective use of the excess capacity and the additional throughput is much more than the actual costs that can be saved by downsizing, then the conflict disappears.

Unused capacity can and should be viewed as an opportunity, not as a liability. First, it helps protect the basic customer base. Beyond that, if we really do have excess capacity, we should try to capitalize on it to generate more throughput. If there are always excess tables at the G-Roy Hotel restaurant, then the management should first determine how it might attract more customers to the restaurant. Roy might consider attracting guests from the other hotels or travelers who are only passing through. The priority for improvement is first to increase throughput. Only after options to increase it are exhausted should you try to reduce inventory (investment) or operating expenses. Recognizing the absolute need for excess capacity in the most resources, while not fully eliminating it, moves the conflict to a different level altogether. The conflict is not reducing the capacity to its minimum level as calculated on paper but how much excess should we provide. It is important to understand that having somewhat more than the absolute minimum protective capacity is preferable to having less than the minimum, especially when that minimum level may not be easy to determine precisely. Having more capacity might cost more, but it enhances efforts to increase throughput. Having less capacity than is needed for good subordination will eventually cause actual loss of throughput, now and in the future. But let us not go overboard with excess capacity, either. Effective subordination should lead us to a reasonable amount of excess capacity that will help to increase our business.

So, What Should Roy Do?

Roy's challenge is to maintain effective subordination to the major constraint: the market demand. If the lack of capacity prevents the G-Roy Hotel from providing good enough service to all its guests, then the management should temporarily plan for fewer guests at the hotel.

That may seem like a very drastic move. It seems crazy not to book rooms that are available. But to rebuild the hotel's reputation the first obligation is to ensure appropriate service. That requires more capacity. Until more capacity is installed and fully functional, the number of guests should match the current ability to provide quality service.

The number of front desk clerks and cleaners can be increased immediately. Turning one of the convention halls into a restaurant (at least to create more seating) may be possible in a short time. Building a new elevator may take longer. Until that is solved, the capacity of the elevators at the peak times is the internal constraint that dictates how many rooms can be offered to guests. In reality, we know that once customers are lost through dissatisfaction, they become skeptical and are not likely to come back very easily. So, in all probability, Roy will also need some new actions to restore the hotel's reputation or arouse enough curiosity among former customers to bring them back. A public declaration of service might be a good idea.

It is not the intent in this case study analysis to construct complete solutions. Rather, its purpose is to show the direction in which the solution lies and to establish the relevancy of TOC.

My Recommended Way to Analyze a Case Study— Real or Virtual

When you face any problem situation, real or fictional, you are in a position similar to that of a consultant to a new client. The consultant has to analyze the situation without the intuition and in-house knowledge that the management and employees of that organization have. The consultant, much like the reader of this book, has some advantages that can offset the lack of intuition. First, the consultant can set himself aside from the emotional issues and beliefs inherent within the organization. An expression in Hebrew says, "A casual visitor sees all the flaws." External objectivity is of utmost importance in revealing hidden assumptions that might lead to severe policy constraints. Ideally, I would like to enable managers to analyze their own performance with much less distortion, combining both the intuition and knowledge from inside the organization with the objectivity and more generic knowledge of an outsider.

TOC provides a knowledge base that is very helpful in raising key questions. Solving problems takes much less time and effort than what happens when those questions are not clearly raised. The following questions are valuable when you need to analyze the current state of an organization. When you need to tackle the problems of a part of the organization or a more specific problem, you will need to generate somewhat different questions.

One key question when considering a specific organization: Does the organization have a clear and agreed upon goal? In some organizations the goal is clear. In others that is not the case. The answer is central to the identification of the constraint(s).

The next key question: What is blocking the organization from achieving more of the goal? I've already expressed my position that the demand is always a constraint. Still, the question is whether the current market demand is constrained because of an internal capacity constraint. If so, the organization cannot serve more demand than it currently does.

So, another leading questions is: Can the organization satisfy more customers and/or more sales? This last question requires further discussion. Suppose the organization is capacity-constrained for certain periods of time, or constrained for a certain line of products, but could serve more customers at other times. The preceding question should be asked regularly—every time the environment changes and for any product currently sold. The question also assumes that the organizational structure, resources, expertise, and procedures remain the same. Should those change, the questions should also be asked.

Suppose the answer to the last question is yes. The organization is capable of serving more customers at any time and for any product sold. That implies that the only constraint to improved performance is a limited market. In this case, the next inquiry must help us understand why the market is constrained. Here we would look for a policy constraint that serves to limit the demand. A few possible directions to look for include the following:

1. Ineffective subordination to the market requirements.
2. Erroneous cost determination (leading to higher than necessary price quotations or refraining from selling).
3. The inability to recognize a market opportunity because of inertia and misconception of the customers' needs. Inertia means sticking to old thinking paradigms—not noticing that the environment changed and those paradigms are not valid any more.

When the market demand is internally constrained, similar questions should be raised: Is the identity of the capacity constraint clear? If not, look into the subordination processes and the measurement system. As every process has its own set of decision rules, what objective the process subordinates to can be clarified. For instance, efficiencies are a common objective to subordinate to. Hence, the decisions on the floor are oriented to reduce setup time and all other non-added value time. This objective for subordination is claimed by TOC to be wrong. That is because most resources should have excess capacity to start with, and there is no organizational added value in producing large batches to stock up for those resources that are nonconstraints. The real challenge for the nonconstraints is to support the exploitation of the constraint. Individual efficiencies for all resources is a major common policy constraint.

Even when the constraint can easily be identified more questions are raised: Should this really be a constraint? Is it well exploited? Are there good subordination processes to back it up?

Is the first question surprising? In my revision of Step 2, I advise the management to look at the identified constraint and ask themselves: Does it make sense to be constrained by this? If not, let us eliminate it immediately from being a constraint. Having the wrong capacity constraint is a policy constraint that is common in companies that went into TOC without enough understanding of the role of constraints in the success of the organization.

Material constraints are relatively rare. When the lack of material is a root cause for constraining the demand, the subordination processes should usually be checked to determine why material constraints happen and whether there is a simple way to prevent this from happening.

The hierarchy of the leading questions can be presented as a flow diagram in Figure 1.4. Please note that the arrows represent flow, not cause and effect relationships.

Any policy constraint is caused by a thinking paradigm. The perceived problem can be presented as a cloud (conflict resolution diagram) where the paradigm to be challenged is represented by one side of the conflict. Identifying that paradigm opens the way to construct a worthy solution. Any such solution should be verified by a detailed future-reality tree.

The purpose of the future-reality tree is to answer three basic questions:
1. What benefits can we expect from implementing the proposed solution?
2. What negative side effects may emerge when we implement the solution?
3. How can we preclude the development (or at least minimize) these negative side effects?

The logical map is a good way to draw future-reality trees, and the preceding questions are very powerful guidelines for doing it. In TOC terminology,

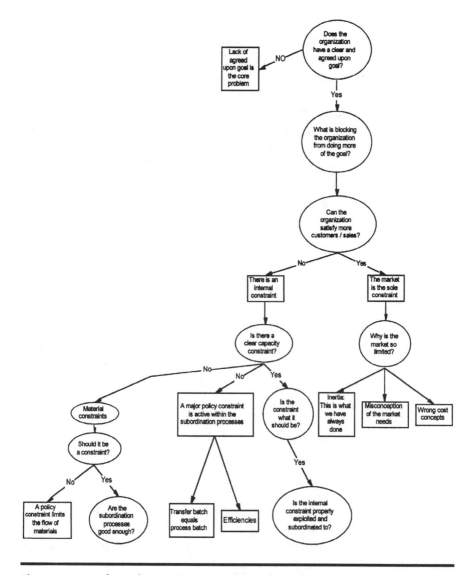

Figure 1.4   A Flow of Questions to Inquire About the State of an Organization

the undesirable side effects are called negative branches. In the simple future-reality tree for mapping the intuitive logic I presented earlier (Figure 1.1), there is just one negative branch that is caused when our flaws are exposed: "We feel humiliated."

Can you draw a future reality tree for the G-Roy hotel? Actions and decisions I have proposed:

1. Enlarging the capacity of the restaurant, elevators, front desk and housekeeping.
2. Until all that additional capacity is functioning, limit the number of guests to an amount that can be adequately supported by the current capacity.
3. Publish a declaration commitment to high quality service.

The negative branches are the need to invest money and maybe give up one of the convention halls. A more serious negative branch might indicate that all of this is too little, too late. Suppose the customers have lost their faith in the hotel and are not inclined to come back very soon. Suppose that the idea of a declaration of commitment to better service is not good enough. What if selling the hotel and changing its name ("Under new management...") is the only thing to do? Would Roy like that idea? Can you visualize the conflict resolution diagram for Roy's conflict? Can you come up with an idea to evaporate that cloud?

With this general approach, go ahead and read the stories that follow. Try to identify the problem and what the main character should do. Try it your own way. Use the TOC tools if you wish. Then read my analysis and decide for yourself whether it is as convincing as yours. If your analysis is very different from mine, try to speculate why. Did I make a hidden assumption that took me in a different direction? Did you? I believe this is learning. And I also hope you have some fun doing it.

2 When Things Are Going Well, Why Change?*

The first story naturally deals with the question of when a company needs to look for improvements. The new management philosophies treat this question almost as a rhetorical one. "Of course every company needs to improve otherwise the competition would wipe it out." This story also tries to highlight the other side of the coin. Certainly the question of when an "improvement" is really an improvement or merely a change emerges. The discussion between the father and the son is, in my opinion, not as trivial as it may look at first glance.

When Things Are Going Well, Why Change?—Case

January 1, 1998. George calls for an annual general meeting of all the employees of The Fast Office. Everybody knows that George will use this formal celebration to present his 32-year-old son, Steve, as the new Managing Director of The Fast Office.

George founded The Fast Office about 20 years ago, when he retired from the army as a colonel. The idea, which he has carefully developed and nurtured for the last 10 years, was to supply office supplies by phone and a delivery service. In those days in this state, the idea was new. George invested

---

* Taken from *Status, The Magazine for Management*, Narkis Weinberg, Ed., 1997. With permission.

all his retirement funds into the project and was well-rewarded. The Fast Office is still the leading company in its field in the area.

George opened the meeting with the following words:

"Dear workers, I am 70 years old. Twenty of these years have been spent in this wonderful organization called The Fast Office. I am proud to hand over the reins to my eldest son, a graduate from the Harvard Business School who headed our project of introducing a computerized network. I would like to take this opportunity to review the company's current situation.

"The Fast Office is the leading company in this city in the field of office supplies. In 1996 our profit was $2.1 million before taxes from a turnover of $27 million. Taking into account the fact that there are only 47 employees in the company, we can say that the net profit per worker and the turnover per worker are high by any standard. I am aware that there was a drop in sales of 7 percent during 1997, but I still believe that the bottom line will show a nice profit for this year, too.

"In mid-1996 we introduced the Internet communication system. A client can contact us through the computer, get our computerized catalog on his screen, and place an order. I am happy to inform you that as of this date, 40 percent of our sales are done this way, and in this area we are a generation ahead of our competitors. But our main strength lies in our warehouses and in our sophisticated purchasing planning. We commit ourselves to large purchases from the suppliers, thus obtaining substantial discounts and better terms of credit than the competition.

"Our delivery staff works properly and we manage to meet the challenge to supply within 24 hours in 95 percent of the cases.

"To find out how our clients feel about our service, we recently conducted a survey. We found out that most of our clients are satisfied with the service and with the delivery of goods. I would like to stress that we received especially high marks concerning the variety of goods we carry. I happen to think that this particular point is a crucial one in business. Clients like to include in one order all the products they need from the manufacturers they are used to. During the last month I requested the secretarial staff to list all the requests for items we do not carry. The list held only 12 items. And you know we have more than 5,500 orders a month!

"I will be the first to admit that clients also have complaints. People like to complain. We all know about it. Our Customer Complaints Department, managed by Rita, claims that 40 percent of the complaints are unfounded. True, some of them are justified. For example, we received a complaint from David Perlinsky, a personal friend of mine and the managing director of the

Pharm-More chain. He complained that the bimonthly delivery of computer paper contained several damaged packages. First I ordered the complete, new delivery immediately, then I personally headed an investigation into what had happened.

"Apparently there had been an unusually strong storm back in November 1997, and some water had gotten into a corner of the warehouse. I am really angry with Jack for not letting me know immediately about the incident. Jack threw out all the boxes that were obviously wet, but he did not inspect the remaining boxes thoroughly. This is not good enough. I expect a higher degree of responsibility from each employee. Had I been aware of the problem, Jack, his two assistants, and I would have conducted a thorough check of all the products that are sensitive to water no matter how long it would have taken.

"I agree, the company needs to improve. But for me, that means each and every one of you has to be more responsible. If everyone tries harder, we won't have to introduce confusing changes every other day. I have seen what happened in Pharm-More as a result of all the "improvement processes" they have introduced. Their workers are so confused with the "service approach" that they don't know exactly what they have to do. They answer the phone very nicely and politely and ask just like robots: "How can I help you?" But when I asked how many packages of paper were spoiled so that I could send a new shipment, they were lost. I think it's stupid. I also can't remember hearing that their business has improved. You tell me: What's the point of improvement if business doesn't improve?

"Last week the managing director of Furniture For The Manager told me they had been ISO-certified two years ago. Big deal! I returned one of their desks that had an uneven finish. And there was a mistake in the bill. Today they think ISO is the solution for everything. The only solution I know is that we all try to make The Fast Office a reliable and profitable company. Even without any improvement processes, today we get fewer complaints than before. I keep track of this subject because it interests me. I want the clients to be satisfied so they will return. In 1997, we got 10 percent fewer complaint letters. I count each complaint, even those that aren't worth dealing with.

"To be short and sweet, I think The Fast Office is an excellent company, first place in the market. I think it is time we expand to the northern part of the state. There is potential there. But I will leave the decision to Steve.

"I would like to end this speech with a few words concerning our relationship with our suppliers. People say that I am a difficult person. You know what? I accept that as a compliment. Business is business. I am ready to purchase a large amount at a discount of 7 percent, order a shipment of four months sales, and store it at our place. All sorts of advisers, who think they know everything, tell me that the price of the storage is higher than the discount. Let them study arithmetic. I have made my calculations. Even an incident such as the one that happened with the paper is still worth the discount. I will tell you something else. What I keep in my warehouse can be sent within two hours at a client's request. If it had to come from the supplier, it would take at least a week to reach the client. We work with lots of suppliers, and most of them don't meet our requirements. That's the world we live in, and we have to learn how to deal with it. If what the client wants is this supplier's product, that is exactly what The Fast Office is going to deliver. That is the reason I keep all the products available. We also make a profit on it. And if you doubt it, you can check our financial reports.

"Let me tell you, dear workers, if I had been a bad managing director, The Fast Office would have been losing money. Then, dear workers, you wouldn't be sitting here smiling. You would have been busy polishing your resumes. I am proud of the fact that I have never fired a worker who has been with us for more than two months.

"The reason is that I have managed to build a good business. Otherwise, with all the good intentions, it wouldn't have worked".

When George finished, all the workers applauded. With a wide gesture he invited his son to address the audience. Steve stood up and started speaking.

"The Fast Office is an outstanding company. For this, we owe thanks to my father, who built it with his bare hands and brought it, together with you, to where we are today. But a company that does not improve continuously is left behind. Only a year and a half ago we heard about the opening of Office-Mart. Rumors have it that they are losing money, but they are harming our customer base. Today, customers have a choice. It is a good thing we introduced our computer network system in time. As my father said, sales by computer today make up 40 percent of the general sales. But we have to do many other things as well.

"Actually, we have to improve and become more efficient in all our internal procedures. The problem we had with the delivery for Pharm-More is not the only one. There has been other damage to the goods we have in storage. A few months ago there was a case with printing ribbons that had just dried up. The client sent them back and we replaced them, but our customers demand the products to be faultless. We need to improve the management of the warehouse, perform proper inspections on goods delivered to us, and sample products that remain for more than a month on our shelves. I also believe that we can decrease the amount of stock we hold in our warehouses.

"I disagree with my father about working with many suppliers. We have to make our relationships with suppliers more efficient, and that's a good reason to work with a smaller number of them. Then we can demand a higher quality in the products and speed in the delivery dates. The bookkeeping would become easier, too, if we worked with a smaller number of suppliers. Mary, the bookkeeper spends a whole week every month just arranging the suppliers' bills and matching them to the deliveries. Most of our clients do not demand a specific supplier, they want paper, diskettes, paper clips, or ink. The manufacturer is not important; they just want the best quality. We could explain to our customers that we work only with suppliers we trust.

"We do not manage the delivery staff efficiently either. Most companies that deliver supplies request a minimum value per delivery. For us, an order for $15 is an order. This causes some offices to ask for a delivery every couple of days. I am ready to supply this service on the condition that the minimum value of each delivery would be $50. There are also problems in finding the addresses. Sometimes the delivery person spends half an hour or more looking for an address. The process can and should be made more efficient. We must find ways to plan an optimal route for the delivery person. We could also figure out the right number of deliveries per round. That will make for a better use of the delivery people's time. I checked with companies in the U.S. and found out that the turnover per delivery person is 40 percent higher than ours.

"Now, about work procedures. I am truly amazed at the small number of problems even though we have no clear protocol stating the rules of receiving an order, billing an order, or how to inspect each shipment before it leaves our premises. I want us to become certified by ISO. This is important from the marketing point of view, and it helps to improve procedures and avoid expensive mistakes.

"Within this framework, we will redefine the rules of our communication by telephone. We have to remember that about 5 percent of our deliveries are returned because of 'misunderstandings,' meaning, the customer claims he ordered something else. Those returns cost a lot of money and upset the clients. I pulled out some of the forms that belong to clients who claimed they received something different from what they had ordered. There is a mismatch between the description of what the client wanted and the catalog number.

"The client wants, for example, wide fanfold paper, but he writes down the catalog number of the narrow one. The clerk who got his order on the phone should have pointed out the mistake. We have an additional problem here. Since the order clerks don't sign the forms, I don't know who made the mistake. Those mistakes may happen in a computerized system, too, but in that case the client marks the item he wants, so he is the one who makes the mistake and there is nothing we can do.

"The warehouse staff who takes care of the packaging needs to become more efficient, too. The delivery people complain that the packaging is not strong enough. But the main problem is that the delivery person has to wait until the guys at the warehouse finish preparing a couple of additional packages. You see? If there are no clear procedures on how many packages one person must deliver, it leads to a big waste of time. We also have the problem of "faulty deliveries," meaning, one item is missing. Either it was forgotten or it wasn't available for one reason or another. Every time such a delivery leaves our premises, it means there will be another one the following day. If this doesn't mean wasting time and money, then I don't know what does.

"But the greatest change will be in the style of management. From now on, a management team will manage The Fast Office. I am bringing with me a good friend and a joint manager, Aaron Kurtz. Our idea is to expand by becoming sole distributors of certain office supplies. We'll sell these products to all the other office suppliers in the state. This will be Aaron's domain, but we will cooperate in any executive decision. We expect that this kind of mutual decision procedure between Aaron and myself will reflect on other managers as well. According to our empowerment scheme, we shall also give more responsibility to the other managers. Rita will become our marketing director, the only one responsible for the communication system with the clients. Jack will be the purchasing director, and Schlesinger will be the operation manager. He will be in charge of the computer system, organizing the deliveries according to the orders, and the quality inspections.

"Each Thursday at 18:00 hours there will be a management meeting of the five of us to take care of anything that has come up. Once a month Josh Rawle, the company's accountant will take part in the meeting. I regard him as an economic counselor for the company.

"Today a new way opens for The Fast Office—we will become better and larger in order to remain number one in the state.

Steve sat down. George looked very upset, but Steve was his son and he had to live with his decisions. The disagreement between father and son did not start at that gathering. Everybody remembered how George had shouted when, two years before, while planning the computerized system, Steve wanted to introduce only 20 percent of the existing suppliers. They were not used to hearing George shout. Until Steve came back from Harvard, no worker had dared to argue with George. George, in turn, never raised his voice, not even when he was upset with a worker. The authority that he demonstrated was enough to make every worker obey him. When he and Steve had the argument, his booming voice could be heard all over the building: "First let us see how you build the whole system properly. Then, when the system works as it should, I will transfer the business to you and you can handle it as you wish." Today George transfers the management of the company to his son, but the argument still goes on. The workers, confused and worried, ask themselves who is right and what should be done.

When Things Are Going Well, Why Change?—An Analysis

The debate between George and his son, Steve, centers around two issues. The first is the constant need to improve every aspect of the business. The second, a more subtle one, is about the value of new ideas about management that have emerged in the last 15 years. When we delve deeply enough into the two issues, we will find the need to focus on the most significant problems as a key message.

Let us express the first issue using the tool of the "cloud" or conflict resolution diagram. In the present conflict, both George and Steve strive to maintain a successful business, meaning, to make money now as well as in the future. How come they diverge so much when searching for ways to accomplish that objective?

Figure 2.1 shows how I interpret the conflict.

In my opinion, this is a generic conflict in many businesses. Introducing changes continuously seems both necessary to maintain a good business and, at the same time, as damaging the same objective. To better understand why an entity can be both beneficial and damaging, we need to look at the necessary conditions (Figure 2.1):

A: This is the mutual objective of both.
   To achieve this, it is claimed that we *must* have both B and C.
B: This entity expresses a well-accepted argument that any successful organization must be constantly improved. It is certainly in line with Steve's views.
C: This is George's claim. He is the one with the experience and proven successes.
D: To achieve B (process of ongoing improvement) we need to continuously improve all internal procedures and become more efficient. Steve's speech reflects his position on this.
D′: To maintain the advantages, one has to refrain from frequent changes. This is based on George's statement to the employees.

It is clear that there is a conflict between D and D′. One demands a lot of changes; the other discourages them.

The TOC way of dealing with such a conflict is to reveal the basic assumptions behind the claims. The arrows that connect entities all imply some basic

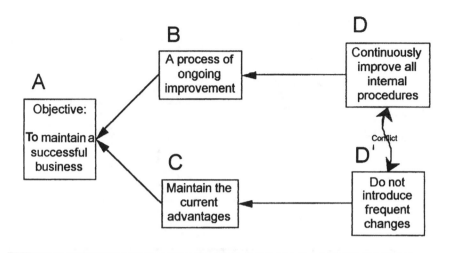

Figure 2.1   The Conflicting Approaches of George and Steve

assumptions. To reveal those assumptions, one should try to think in what environment the logical arrow will *not* be valid. This way may create some new alternatives to eliminate the conflict.

Let us look at the A-B arrow. When does a successful business *not* need a process of ongoing improvement?

For instance, when the environment does not change. In such a case, a successful business will stay successful. So, in claiming that in order to maintain a successful business there is a need for a process of ongoing improvement, it is assumed that the environment changes continually. Another example is in a totalitarian government in which the business belongs to the dictator. That example leads to recognizing the assumption of operating in a free market for the ongoing process to be necessary for maintaining a successful business.

Regarding B-D, we usually assume that ongoing improvement can be achieved only by improving all or most of the internal procedures, otherwise the global improvement will not be significant. In other words, every individual improvement has some effect but the significant effect is the sum of many improvements throughout the company.

The TOC philosophy challenges that assumption, arguing that improving very few procedures at a time can have great impact on the bottom line.

Regarding D' as a necessary condition for C, an underlying assumption is that many changes inevitably impact those areas that perform quite well relative to the business objectives. This basic assumption can be challenged.

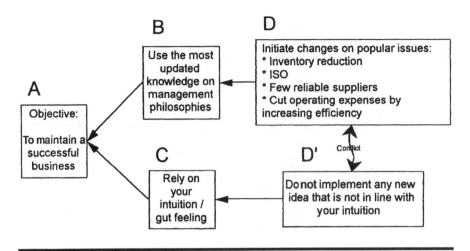

Figure 2.2   A Generic Conflict with the New Management Philosophies

If the changes are to be directed only to those areas that prevent the business from additional success, then the better areas will not be adversely affected.

What about the dispute over the new management concepts? The next cloud (Figure 2.2) illustrates the conflict.

A noted TOC author, Bill Dettmer, has suggested that the B-D and C-D′ links offer the better chances of evaporating the cloud.

The basic assumption behind the B-D arrow is that using the most updated knowledge means implementing it as is. That assumption is challenged by claiming that although the new ideas should be considered, they should be carefully checked and tailored. Inventory reduction is a good idea in general, but one should ask about the role of inventory in this particular case and then try to see whether the inventory levels are excessive or not. The TOC philosophy challenges Steve's preoccupation with "efficiency." That, of course, does not prove TOC is right, but, at least it is worth exploring.

The essence of the injection for B-D is that those actions are not necessarily good. They certainly are in conflict with George's intuition. Let us check all those ideas more thoroughly.

How are we going to decide what idea is worth implementing? George provided a generic guideline: "What's the point of improvement if business doesn't improve?"

Put in a positive way, we should look for the improvements that will clearly impact the business. The lack of improvement in these areas prevents the company from doing better business. Hence, we should look for those areas that limit the company from making more money.

This is TOC's most important question: What is blocking the organization from achieving more of the goal? In this case, the goal is definitely to make more money.

An obvious possible answer is: The company can make more money if the volume of sales is larger but expenses remain the same. Such a statement is valid only when more sales mean more profits. There are few cases where this might not be true. One might be a situation in which the additional sales also generate significant additional expenses. Those extra costs might be greater than the increase in turnover. Another situation is when the extra amount of sales cause the service level to drop, discouraging future sales.

The Fast Office sells items it buys from its suppliers. We can assume it sells each item at a price that is higher than its own suppliers' price. All the other expenses of the company are not directly proportional to the sales volume. If we succeed in generating more sales but we do not hire more people and we store the inventory in the company's own warehouse, we can assume more profits are going to be generated.

The second situation, harming the service level, is one to be more concerned about. If one or more of the current resources is working close to their maximum capacity, there is a risk that additional sales could jeopardize the current service level.

In the TOC terminology the question is: What are the constraint(s) of The Fast Office?

According to the TOC methodology, an organization is constrained by very few variables at one time. When a company cannot satisfy all the demands for its products or services, it has an internal constraint somewhere within the organization.

Does The Fast Office have a resource that is "almost a bottleneck?" Such resources leave their footprints on the behavior of the company. Nothing like that is evident in the story. On the contrary. In 1997 the sales volume dropped 7 percent. That means The Fast Office can supply at least 7 percent more sales. So, no internal capacity constraint is limiting the sales.

Certainly the warehouse capacity is not a constraint. Had the warehouse been full almost to its limit, George would not buy a lot of stock on discount. Had the delivery people been loaded to their capacity, they would have made a lot of mistakes trying to hasten the packaging operations. They would have returned deliveries when they had just a slight difficulty locating the address, leaving certain deliveries to sit for a long time because it was not convenient to go that way and complaining a lot about the pressure. It is impossible to achieve 95 percent of on-time deliveries with fully loaded delivery personnel.

Based on those clues, we can deduce that The Fast Office can supply more goods with the same quality of service. The constraint is the market demand! The problem is that existing customers are leaving faster than new customers are coming in. That is where the immediate improvement efforts should focus.

A reasonable hypothesis explaining the loss of sales is that delivering within 24 hours 95 percent of the time, as good as it may look, is not good enough. Maybe it was good before the competition emerged, but not anymore. Having merchandise returned is bad, not because of the extra work and extra expenses, but because customers switch to the competition (even though it may be no better).

How can The Fast Office improve its service level? In the TOC terminology, it means instituting better subordination processes. When it is so important to satisfy the customers, every procedure in the company should ensure the customers' satisfaction. George should not blame the workers. Blaming others does not improve any business; implementing appropriate measurements does. When the measurements count faulty shipments and assign them to the responsible worker, the workers will try hard to prevent faulty shipments from happening. When the warehouse manager feels that throwing away wet paper, because of some defect in the warehouse ceiling, is a loss directed to him, he may ignore those packages that "look" alright. A good subordination to the market constraint should ensure there are no returns of damaged paper. That is more important than saving the costs of the paper. The measurements should direct the warehouse manager to set the priorities right. A better subordination process would be to store the paper at a certain place for further inspection and not use it unless it is fully approved.

Because the market demand is the ultimate system constraint, more ways to **exploit** or **elevate** it should be considered. Improving the subordination processes helps to exploit the current market demand. For instance, having fewer returns of merchandise means not giving the customers any reason to go to the competition. That improvement in the subordination leads to better exploitation of the current constraint. Although there is still room to improve the subordination processes, it seems that the processes are not too bad even now, and it is time to look for ways to elevate the constraint by finding new market demands.

There are several ways to bring in more sales. Maybe there are other products that are not normally considered "office supplies" that people would like to have delivered to their home or office. One way to generate more sales is to take advantage of the excess capability of the system. The core competence of The Fast Office is its ability to store a variety of items and deliver them to the home or office. An expansion of the product line may cause the need to find the new customers, but the generic knowledge exists within the company. Partnering with book, computer, and hardware stores in a venture to enable them to make deliveries and to store inventory at The Fast Office is also a possible way to look forward.

Please note, expanding the sales without increasing the manpower until an internal resource is on the verge of becoming a bottleneck is a clear straightforward way to improve the business. When one does it carefully, it is a *much* more promising way than trying to cut operating expenses, which prevents the business from growing.

How does all that fit in with what George and Steve have in mind?

George does not recognize the need to re-evaluate the state of the company and improve the weakest links. Improving the service level, either by better procedures or by more appropriate measurements, is necessary. Thinking about new ways to promote sales is another avenue.

Steve has many ideas. His basic need is to focus on the most significant areas. Some of his suggestions may even be harmful. For instance, what is the point of improving the efficiency of the delivery people? You can save expenses that way only if you let some of them go. How certain can one be that such a move will not cause problems in the response time? The fewer people you have, the less flexible you are. Trying to optimize the delivery routes will not enhance the response time and will not save any costs (except some gas, which certainly is not significant). This is a good example of the conflict between efficiency and flexibility. The more efficient you are, the more difficult it is to respond to any change and new need that our uncertain reality imposes on us. When it is difficult to respond to a last minute change, it means poor subordination because the subordination logic demands a response from you to any new need of the constraint.

Should The Fast Office introduce a minimum value to the total sum of a sale? It seems that the total sales will go down. Certainly some customers do not want to buy more than they need to. So they would go either to a retail store or to a competitor. Can The Fast Office allow this to happen? Will the operating expenses go down when the number of shipments goes down? Only by laying off people. Is it worth it? Will the customers be satisfied?

The lesson to be learned is that you need to focus on the most significant issues. You judge what is significant by its impact on the goal of the company. This analysis uses the TOC guidelines to look for the areas where improvements are most promising. Other stories will use more elaborate techniques.

3 | Let There Be Light*

The next story takes us to a familiar territory for TOC fans. A small manufacturing organization with a lot of well-known problems that the classical TOC has dealt with. But beware of inertia. I don't think this is an easy case at all, especially to TOC practitioners who feel they have been there. First of all, as in the first story, not all the facts are clearly presented. All you have is a general statement of the situation and some citations of what the managers say. Still, if the situation looks confusing, go back to the basics. It is all there.

Let There Be Light—Case

The tale of "Let There Be Light" begins with two brothers, each harboring ambitions in a different field. Eric, a graduate of the Bezalel School of Art in Jerusalem, considered himself an artist and designer. In 1994, he was working as a senior graphic artist in an advertising office, having failed to persuade industrial enterprises that they needed his special talent. Owen, his younger brother, had studied economics and management and had worked in a firm of brokers for about 10 years. He was looking for a way to establish and manage a business of his own.

The idea for the Let There Be Light Company emerged when Owen received a halogen desk lamp as a birthday present from his colleagues. When Eric popped into his brother's office, he took one look at the lamp and began criticizing the cumbersome and inelegant design. Eric did not really mean

---

* Taken from *Status, The Magazine for Management,* Narkis Weinberg, Ed., 1997. With permission.

to hurt the feelings of Owen's colleagues who were in the room at the time. To reduce the embarrassment, Owen suggested commissioning the design of a desk lamp from Eric. "Show us all that you can do much better than that," Owen said. Eric enthusiastically accepted the challenge and even volunteered to have his design made. He found a metal workshop willing to make the lamp for very little money. About a week after Owen's birthday, Eric marched into Owen's office and festively placed his lamp beside the other one. Owen looked at it, and his heart missed a beat. He suddenly realized that he had found what he had been looking for for so long. This was a very impressive piece of art, and it gave the whole room a kind of sophisticated status that he assumed many would like. His excitement grew when his colleagues in the brokerage loved the lamp to the extent that they ordered 10 similar lamps from Eric at 10 times the production price.

Let There Be Light was established within four months. Owen obtained financing, and together the brothers purchased the workshop and hired six production workers. Eric designed four basic models, each of which was offered in three different colors.

The company's initial success was astounding. According to Owen's calculations, about 60 percent of the stores he approached immediately placed orders. He went from store to store promoting the lamps, while Eric stayed at the workshop and organized production. Within two months, they accumulated six months of production work. Owen, seeing that they would be unable to meet their promised two-week delivery deadline, was willing to increase his production investment. The brothers bought extra property adjacent to the workshop and hired additional production workers. They purchased more welding machines and a heat-treat oven. Owen also hired a marketing expert to help expand their market. This man managed to sell the idea to the Sharper Image chain, which alone doubled the number of lamps ordered.

During the whole of the first year, Let There Be Light was significantly late with deliveries. Notwithstanding, the clientele expanded. Eric designed some additional models and demand rose further. In early 1995, the company moved to an entire floor of an industrial building. The number of employees rose to 25, including an experienced operations manager. Toward the end of the year, demand stabilized. Let There Be Light promised delivery to stores and marketing chains within a month but often missed deadlines. Sam, the operations manager, complained that the large number of models prevented him from streamlining production.

At the beginning of 1996, Let There Be Light was producing about 30 models of desk lamps in various colors. In March 1996, Owen and Eric decided to reduce the number of models to 10. Eric insisted on maintaining the range of colors, so there were actually 50 different products. That policy meant that each time a new design was introduced, a previous model was discontinued.

Final figures for 1996 showed a 6 percent decline in sales compared to 1995. Owen and the marketing manager decided to expand the company's market by proposing a range of models for domestic furniture companies that wanted to include them in their stores. Eric designed those models so that they would not compete directly with the most expensive ones. Yet despite this fairly successful measure, total sales in 1997 continued their downward trend, reaching only 80 percent of the 1996 total. The bottom line was that the profit margin in 1997 was a mere 1 percent.

In January 1998, the managers of Let There Be Light met. The subject of their meeting was defined as "Shaping the future of Let There Be Light." Owen asked each manager to prepare a report on one aspect of the company as a basis for a discussion on its overall state.

Kathy, the marketing manager, opened with a presentation of the company's products, color catalog, order-lodging process, and distribution of products throughout the state. Owen interrupted her, saying sales were on the way down and there was no point in disguising the bitter truth. Kathy said that although the market was favorable, the company's operations were failing to respond to the market's needs. Sam immediately quipped that the marketing department was constantly demanding products it could not subsequently sell.

Here Eric intervened, saying that Kathy should be given a chance to complete her marketing description. Modest, introverted, and much liked and admired by the company's employees, Eric was always the one, in his quiet way, to placate the other managers when they started arguing. He did exactly that now.

Kathy raised the following points:

- Let There Be Light was still considered a producer of exclusive desk lamps. As recently as November 1997, a very enthusiastic article about Eric's designs had been published in a leading national design magazine.
- There was a steady demand for new designs. Most of Eric's new designs sold extremely well for three months.
- Customers did not tend to leave, but as time went on they ordered fewer lamps.

- Goods were being delivered to the stores within approximately 45 days from the date of order. About 40 percent of deliveries took less than three weeks to be delivered, but 33 percent were late, sometimes by more than two months.
- Previous models were stored in a depot of finished products. Although a market for these models existed, Kathy did not want to sell them at reduced prices for fear of damaging sales of newer models.
- The sales to the furniture companies were proving problematic because the companies were interested in adding their own touch to the basic designs, for example by choosing nonstandardized colors and/or combining models. They also required special engraving. The minimum delivery time for such orders was fixed at three months.

Elie Schremm, the controller, interrupted Kathy by saying he intended to talk about orders incurring a loss, and the orders for the furniture companies were a good example. In Elie's mind those sales lost money because of all the extra work and the reduced price. At this point, Owen said he was in favor of those orders, since they were large and accounted for about 20 percent of all sales. Elie was not convinced, but he let Kathy continue.

- Special orders (10 to 15 units) of exclusive designs were also available to senior managers. On the controller's recommendation, Owen set the price for that type of order at five times the price of the regular exclusive models. As of the beginning of 1996, only two such orders had been placed. Kathy claimed that consumers were being frightened off by the high prices and that at three times the regular price, many more orders would be received.

Sam, the operations manager, raised the following points:

- Production planning was based on marketing predictions for six months in advance. In practice, however, the market was clearly in constant flux and required immediate adjustments.
- The turnover of models was too rapid, with most models being produced for a period of only four to six months.
- The factory required a lot of manual work, especially on the body of the lamps because they had to be bent and plied in a way that made automated machining difficult.

- Changing models involved between half an hour and an hour setup time and were a waste of raw materials.
- Changing colors also meant a waste of setup time.
- Since the oven consumed an extremely high amount of energy, they had to take full advantage of its surface area (70 to 90 lamps).
- In early 1997, some complaints were received about colors fading after several months. It became clear that certain colors (about 15 percent of all lamps) needed to stay twice as long in the oven. Since such exposure to heat risked destroying the other colors, the lamps had to be sorted according to color, which influenced production planning and delayed delivery by a another fortnight.

Elie started by saying he thought Let There Be Light was at a crossroads. "If the status quo continues, we will lose about half a million dollars in 1998," he said. "Operational expenses are increasing, whereas the market, far from expanding, is becoming smaller."

His main points follow:

- Direct wages accounted for over 45 percent of Let There Be Lights' total expenditure. That included 6 percent for overtime hours, a figure that had been constant for two years.
- The company's employees were significantly better remunerated than similarly qualified employees elsewhere. Eric's explanation was that he needed particularly highly skilled workers. He required those working on the lamp bodies to have studied industrial design even though they did not design anything at the factory. Since these workers were so highly paid, the other workers also received good wages.
- The value of the lamps stored in the depot of finished products was 23 percent of the total sales in 1997.
- Despite the relatively high cost of labor, many orders required extra labor with no corresponding compensation in the consumer price, resulting in losses to the company.
- The factory's energy consumption constituted about 10 percent of operational costs, far above average for this branch.
- The cost of raw materials was fairly low (18 percent) compared to sales.

After hearing this apparently gloomy description of the company's state of affairs, Owen mentioned what he considered a potential watershed. A month before, a sales agent had offered his services in finding European clients. Owen had given the agent the catalogs of lamps produced over the past three years and two recently designed lamps not yet being produced. That morning, before the meeting, the agent had called and said he could immediately sign on a delivery of lamps to a French distribution chain. The number of lamps in the order was approximately 30 percent of the total number of lamps produced in 1997. The company wanted the lamps to reach France within 21 days, and Let There Be Light would receive less than 80 percent of the local market price per unit after the deduction of extra shipping costs.

This opportunity evoked mixed reactions from those at the meeting. Elie said he wanted to read over the contract carefully but that the price sounded too low. Sam wanted to know whether the same models for export could also be sold at the local market and whether they could receive an accurate forecast two or three months in advance. Kathy said she was familiar with that French chain and that it was known for the high quality of products it distributed. Kathy added that certain remarks had been made that Eric's designs were more European in taste. That might explain why the demand was thinly spread throughout the U.S., while there was higher demand in Canada.

Eric's thoughts were in a different direction; he was in favor of expanding the range of products and producing metal desk clocks. He took out two superb clock models he had made himself. Like the desk lamps, the clocks were intended for managers' desks but were attractive to others as well. In his opinion, current production resources and manpower were sufficient to produce them. Eric also raised the possibility of producing plastic desk lamps and clocks. Direct labor costs would be lower for plastic than for metal, but producing plastic goods would mean investing in a production line.

At this stage the managers started discussing the future of Let There Be Light.

Let There Be Light—An Analysis

Let There Be Light is a gold mine whose owners do not know how to dig for the gold. This is, in my opinion, quite a typical case where something of great value is lost because of the weakness of "regular" competence.

Both the classic TOC and the thinking processes can be used here to identify the core problem and suggest a way to make Let There Be Light an extremely successful company. TOC teaches us to ask the following questions.

What Blocks the Company From Making More Money?

TOC suggests that any organization can be blocked by very few critical factors, called constraints. The formal TOC definition of a constraint: Anything that significantly limits the performance of an organization.

So, the company is blocked by one, maybe two, constraints. What are they? The market demand is certainly a constraint. The company clearly wants to sell more. The controller remarked that some of the products lose money. He suggested selling those products at higher prices or stopping their production. Owen and Kathy objected to that because they assessed that those products would not sell at all if the price were higher, and they felt, that they were still profitable even at the lower price. Those are the characteristics of the current market for the company's products.

What can be done to expand that demand or increase the perceived value in the eyes of more people?

We do not know much about the marketing efforts of Let There Be Light. From what we know, we can think of several things that adversely affect the market:

1. Streamlining the models offered.
2. Long lead times.
3. Unreliable deliveries.
4. Some quality problems.
5. High costs that prevent selling to some segments of the markets.
6. Eric's designs are more "European" in taste.

There is enough evidence to show that the first blocking factor is real. New models sell nicely for three months. If the company introduced more new models, total demand would probably rise, too.

The true impact of long lead times and the unreliability is harder to determine. In the first year, it did not stop sales from growing.

The quality problems are recent, and they affect only a small portion of the market. The real negative effects might occur in the future if those flaws are not fixed fast enough. Right now it does not seem to be a major impact on the current demand.

The fifth item is evident. The French company is available and the management has reservations concerning the cost. Another case is the exclusive designs for senior managers that are offered at too high a price because of the costs associated with them. The last item opens a new direction that is validated by the French offer.

All of the above lead to one conclusion: There are more opportunities in the market than the company is capitalizing on. The reluctance to take advantage of those options stems from some limitations in production. That means some kind of constraint limits the company in exploiting the market potential.

Is There a Capacity Constraint Within the Shop Floor?

That is an easy question to answer. There is definitely *no* capacity constraint within the shop floor. It is enough to know that in 1997 sales were only 80 percent from the previous year's level, which was lower than the 1995 level. That means the same facilities could produce more products. One might assume that some employees were laid off when the drop in sales occurred. If this had been true, it would have been expected that this move be mentioned. Certainly, the controller, who has to estimate the drop in costs, should have told us about it.

How come the drop in sales did not improve response time? It had to be that the shop floor was run according to the traditional production management norms that tried to "better utilize" the production resources, especially the more expensive piece of equipment—the heat-treat oven. For the sake of efficiency, larger process batches were used, based on a six-month forecast. For a product life cycle of four to six months, that was a grave operational mistake. The result was that 23 percent of the previous year's sales were sitting in the finished products storerooms as old model lamps. How efficient was it to produce 23 percent excess stock? How efficient was it to delay a small batch of a certain color until there were enough units to look reasonable to put them in the oven? Something was very wrong there, but those norms are quite common.

The crucial point here is not that much faster times could have been achieved, even though faster response may be needed in the future. It is the identification of a root cause: a management paradigm that is incorrectly applied. The paradigm affects not only the response time but also causes the strive for streamlining the production and giving up some highly profitable market segments.

How do the traditional paradigms conflict with the TOC focusing steps? When the identity of the system constraints is not acknowledged, there is no proper subordination to the constraints. When Sam complained that changeover was wasting time, what could we learn about his objectives? As no capacity constraint existed at that time in the shop floor, why should he have been concerned with the time wasted? The simple subordination scheme should have been to reduce batch sizes to match the actual sales orders. When we do it that way, the customers are getting their orders *much* faster. As a result, the stores are able to stay in closer touch with the market tastes and requirements, promise quick delivery to interested customers who want a particular model or color, and eventually sell more lamps. Shrinking the lead times could reduce the need to rely on a forecast, thus refraining from producing products that may not sell at all.

What is There to Lose From Proper Subordination to the Market?

The only fear is that by planning too many setups, some resources will turn into bottlenecks. In this case, more capacity will be needed and the operational expenses will go up.

There is no question that there is a significant amount of excess capacity that should be viewed as a means to improve the subordination to the market. When that is done, it is going to relieve several of the undesirable effects mentioned during the meeting:

- No need, or at least less need, to produce according to forecast. That will eliminate the cases of unsold finished products.
- Reducing the finished goods inventory will free more capacity to serve more markets once the decision to look for those markets is made.
- The quality problem will vanish, as every color will have the appropriate oven time without having to depend on the size of the batch.
- More variety of models could be launched, which will lead to increased sales.

The other undesirable effects stem from the cost concept. Can any lamp produced by Let There Be Light lose money? By losing I mean that selling it at the current price will result in less profit (or greater loss) than not producing that lamp. The very few items of information are enough to substantiate a claim that this is *not* the case here. Every lamp sold increases the profit of the company!

How can we know that when no detailed cost analysis appears in the story? The general structure of the costs is enough to make an assessment. Only 18 percent of the sales are raw material costs. That is an average. Suppose that for a certain order it may be as high as 30 percent. Add to it the commissions that may be paid to the sales agents (there is no indication of it, but it may be the case). We may come to realize that a maximum of 35 percent of the sale are truly variable expenses, or expenses that are caused by any such sale. All the rest are added value to the company profit. In TOC terminology, every sale adds at least 65 percent of the sales revenue to the total throughput.

As long as every lamp sold gets a price that is larger than the truly variable expenses, that lamp adds to the profit because all the other expenses do not change because of that single lamp. The only exception may be if some amount of overtime, which is paid by the hour, is needed because of that lamp. In most cases, overtime is used to catch up with delays, cover for flaws in the planning, and better compensate the workers. It is not commonly used to add real capacity. Hence, the overtime expenses usually are not included in the calculation of throughput. But had overtime been a real expense driver at Let There Be Light, it would have been mentioned. So, we can assume it is not a real factor to consider.

When do we have to consider the trade-off between different products? Only when producing that particular lamp is at the expense of another lamp. That means we cannot produce both with the current capacity. That situation could happen only when there is an internal constraint in the company. It might occur in the future, but right now every additional lamp is profitable.

The two main root causes for poor profits of Let There Be Light are generated by the same management paradigm. That paradigm assumes that to optimize the performance of the system, each part can act independently of the other and strive to be as efficient as possible. When the interdependencies are properly understood, the gravity of the problem with this idea is revealed. The only way for a manager to control the system and derive fairly stable output from it is by preserving enough flexibility on the vast majority of the parts. That means excess capacity is not waste; it is an absolute necessity.

What should Let There Be Light do? First, the management has to understand the idea of subordination to the constraint. What is immediately derived from such a recognition is that the operational policies need to be changed. The new policies need to be based on the market as the sole constraint, at least in the next short term. Two clear main actions are needed to improve the subordination to the market. First, the production needs to be planned based on firm orders only. Second, batches need to be much smaller than currently used to achieve very fast response to any request coming from the market. Using small batches may require more setup time, but as long as the excess capacity is not exhausted and some degree of flexibility to new demands and changes is maintained, it is worthwhile. Some more actions should be derived from the drum-buffer-rope (DBR) methodology. Drum-buffer-rope is the TOC manufacturing planning methodology, and it is detailed in several TOC books (see the bibliography at the end of this book).

It seems clear that before the change in the operational policies takes place, there is no point in making the agreement with the French distributor. Failure to meet the requirements may be too risky for the future.

Of course, to start such a change, a comprehensive understanding of what it is all about needs to be in place. No local performance measurements are mentioned in the story. If they do exist and people are judged by their "efficiency," the education and management dedication to the operational change are absolutely necessary.

Once the operational system stabilizes, the second wave of taking advantage of the change starts. The obvious move is to strive for more sales. Two precautions should be taken first. An operational control system must be implemented to identify the emergence of an internal constraint. When a company moves to expanding its sales, the risk of committing to too much can be devastating. The control system should warn the management that a certain work center will lose its excess capacity. That is not done by mere calculation of capacity (even though it is good to have this as an approximate prediction).

It has to be based on the stability of the planning. Buffer management, the shop floor control mechanism of TOC, is able to do all of the above and is a real necessity here. The buffer management technique is explained in most books that deal with the drum-buffer-rope methodology.

The other precaution is to predict which work center is going to create the next constraint—and according to this prediction, expand sales. From this story, one can get the impression that the oven is most likely to become the capacity constraint. This, of course, is only a hypothesis that should be carefully checked in reality. What makes this assumption ring true is that it takes a long time for the oven to complete a whole batch. Because the general tendency is to use smaller batches and not wait for the oven to be full, it might turn out to be a constraint. When that happens, the subordination processes should allow filling up of the oven, and it would be desirable to design the product mix according to the throughput (added value) per hour of the oven.

The oven, or another capacity constraint, will emerge as a constraint in a medium time frame. An oven is not likely to limit the performance of such a company for long. An oven usually is not such an expensive resource. Elevating a resource for which the investment is not particularly high is a good move because the potential throughput that can be gained is usually much more than the investment.

Developing a long-term strategy means deciding where the long-term constraint will be. Realizing that the most unique competence in the company is Eric's ability to design products that people are willing to pay high prices for makes it a potential long-term constraint. What are the ramifications of such a realization? We have to be looking for those markets in which Eric's unique ability is especially rewarding. This is where a management vision is needed, a vision that directs the organization to make its most unique feature the future constraint. That also means a good exploitation of that constraint in which the rest of the system is built to subordinate to that exploitation scheme.

Let There Be Light—Using Some of the TP Tools

The above analysis is a combination of the guided questions, the basic concepts and ideas that were introduced by TOC, and some intuitive logic. As we did with the G-Roy Hotel story, let us now try to state the intuitive arguments in the structured way as proposed by the TOC's thinking processes (TP). Here are some sketches for the analysis of Let There be Light (Figure 3.1).

The first step maps the basic arguments used to answer the question: What blocks the company from making more money?

This is an alternative verbalization of the possible causes. This is just the first sketch. The cause and effect between the blocking causes have not been looked into yet. Right now even the proposed cause and effect links are dubious because some of them may not be significant enough. Certainly some causes are not sufficient to cause the effect at the top. But putting those

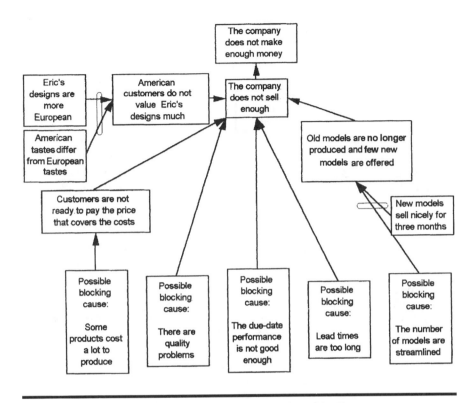

Figure 3.1   Several Possible Blocking Causes to the Company's Performance

on paper is a starting point from which to look for the more meaningful cause and effect relationships.

Looking hard at the proposed connections, only two have enough substance to explain the statement at the top. The due-date performance, the lead time, and the quality problems do not seem to be *the* cause. They might be by-products of some other cause that is more substantial. Figure 3.2 shows still a very crude draft of a more meaningful tree. The emphasis is on two possible causes that look like better candidates for explaining the effect at the top.

Figure 3.2 is not good enough. It is just a temporary stage where I have put something on paper in order to better examine it. The inclusion of temporary drawings here shows that in our (well, at least in my own) attempt to establish a better understanding of the cause and effect, we may include some drawing with partial and even wrong logical arguments. In the search for the core problem, it is vital to put our arguments on paper and carefully scrutinize the logic.

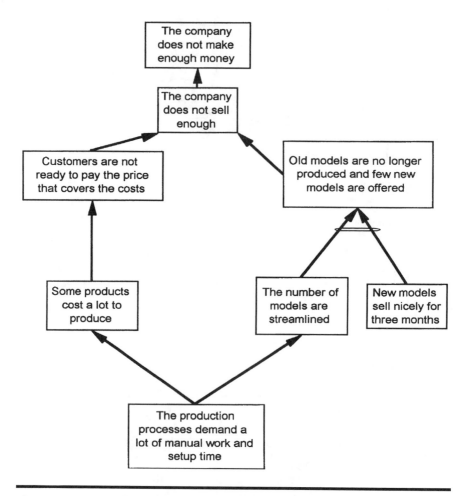

Figure 3.2   A Crude Cause and Effect for Further Development

The statement at the bottom certainly does not look right. Somehow we feel that the high cost attributed to some products, and the policy of streamlining the number of models seems to be connected with the setups and manual work at the production shop floor. Moreover, a core problem needs to be connected with our thinking paradigms. Otherwise, we cannot do anything to really improve. Any real improvement has to be based on something we have failed to notice until now. Hence, we should look for wrong paradigms that limit the whole organization from achieving more of the goal. Although every organization has at least one constraint, it can also always improve itself by finding ways to better exploit, subordinate, and elevate. All of those are achieved by overcoming some "blind spots" in our cause-and-effect understanding.

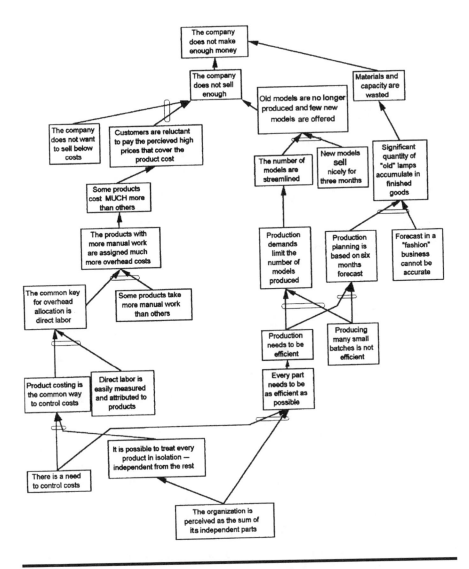

Figure 3.3   A Summary of the Current-Reality Tree

A more sound logical map of the current-reality tree of Let There Be Light is presented in Figure 3.3.

The recognition of the stated core problem, a faulty-thinking paradigm that sees the organization as a collection of independent parts, is central to TOC. All newer management approaches agree that the interdependencies within an organization are very considerable, and an organization cannot be perceived as consisting of independent parts. However, the vast majority of

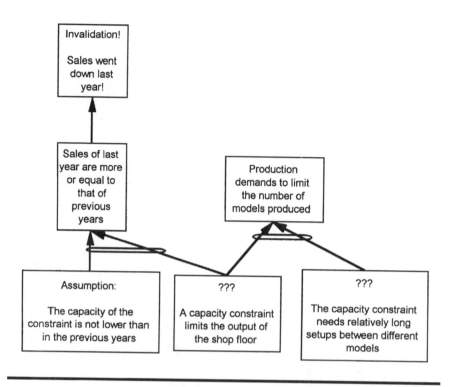

Figure 3.4   Ruling Out the Assumption That a Capacity Constraint Exists

companies behave this way. What else can drive Sam, the operations manager of Let There Be Light, to refrain from subordinating to the market (which is the responsibility of another part of the company) but rather to some internal efficiency norms that do not contribute anything to the company as a whole?

In the search for the core problem, one might ask why I assumed that the effect "Production demands to limit the number of models produced" was caused by the efficiency syndrome and not by a capacity constraint. Certainly an existence of a capacity constraint might have caused a limit imposed on the number of models. That is part of the tree that is not fully explained here. The key guiding question—Is there a capacity constraint within the floor?—directs us to look for alternative causes for the above effect because we can deduce that there is no such constraint due to the following short logical argument that invalidates the claim that it exists. The challenged effects appear with question marks to show they are being tested for their existence (Figure 3.4).

The assumption that no capacity constraint was generated because of decreasing its capacity is a necessary one. However, the story tells us about

A Common Conflict In Management
Looking For The "Best" Compromise

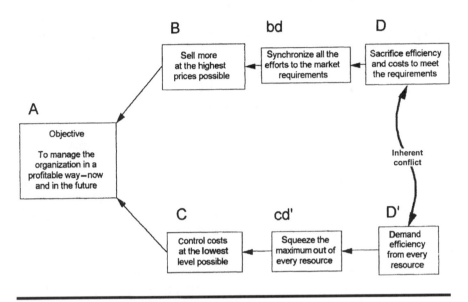

Figure 3.5   A Common Conflict in Management

expanding the capacity throughout the years, not reducing it where it is most needed. So I think this is a fair assumption.

Managing the dependencies in the organization drives the managers to look for synchronization. This is an extremely tough challenge. On the other hand, there is an obvious need to control costs. Controlling the costs is very local in nature and drives the managers to look at the efficiency of each one. That creates a basic conflict.

The steps taken by Owen, as the president and CEO of the small company are to find a compromise between Sam's demand to limit the number of the models and colors and Eric's and Kathy's desire for a wide variety of models. The same thing goes for Elie's complaints about certain products being too costly or their selling price being too low. The channel of exclusive designs for senior executives is blocked by very high selling prices.

Recognizing the need for excess capacity helps to break the assumption drawn here under cd' (Figure 3.5). If you assume that every resource is fully loaded, then any additional work is at the expense of something else. Another assumption is that any additional work carried out by a resource causes expenses. Those assumptions are challenged by TOC. That means sometimes additional work does *not* cause any additional costs. More products can be manufactured with the current resources, and the only out-of-pocket money is for the materials, commissions, and perhaps some energy. The minimum of synchronization that needs to be in place determines the level of excess capacity and the costs. Strict subordination to the market should be the major concern unless there is an internal capacity constraint that is too expensive or difficult to elevate on the spot.

After the main conflict has been evaporated, there might be cases where *some* considerations of efficiencies are needed. When you fully subordinate to the market requirements, a capacity constraint may emerge because its excess capacity is all taken by setups or by time-consuming customer requirements that do not seem to add much value. Machines with a relatively large setup time *may* impose a certain minimum batch policy. The TOC insight is that one should consider merging batches only when those setups actually cause a bottleneck. The significance of the above conflict is mostly evaporated, and clear guidelines can be developed to decide when certain changes in the subordination processes are needed.

Let There Be Light is a story about wrong subordination processes caused by a management paradigm that is wrong but very common. The impact of that particular paradigm is huge! It causes wrong operational policies as well as wrong perception of what is profitable. It also highlights the importance of first recognizing the core problem, then fixing the operational policies, then the marketing and sales policies, and last creating a new vision of the company that would be based on the most precious internal resource the organization has.

4 | A Laboratory for the Repair of Communication Equipment

T he five focusing steps of TOC are very obvious in manufacturing. The following story is about a service organization that is part of a nonprofit one. Are the five focusing steps relevant in this environment? Think about it when reading this story. From a different angle, think about the characteristics of a peak of demand. Most organizations from time to time have this kind of a huge, though temporary, peak load. Can anything be done to relieve the pressure?

A Laboratory for the Repair of Communication Equipment—Case

The Central Communications Laboratory of the Army repairs communication equipment, including an attached electronic warfare device. This is the last year these instruments will be in use. Next year they will be replaced by a new-generation device. Because of the anticipated change, there have not been any new purchases of old-generation equipment during the past two years. The stock of spare parts for them is diminishing.

The current policy is to accept broken instruments from the field units, fix them whenever possible, and return them for further use.

These days activity at the laboratory is especially heavy. On one hand, people are getting ready for the integration of the new-generation instruments and preparing a proper maintenance base. The laboratory has received an advance shipment of 10 new-generation instruments for this purpose. For security reasons, all the activities connected with the new instruments are being held in a separate building called The New Building. The commander in charge, Major Williams, has formed a special group of seven soldiers to isolate themselves in the building, study the structure of the instruments and prepare instructions about the different inspections that should be carried out on the instruments once they are introduced into service.

On the other hand, the current activity centers around the fixing of the old instruments. Additional maintenance staff has not been approved even though the number of instruments that need to be repaired has increased. It now stands at about 120 instruments a week, four times larger than before.

The procedure for receiving faulty instruments is fairly routine. The instruments arrive directly from the unit or from a logistic center. Every two days a driver leaves the laboratory and goes to the logistic center. He returns the instruments that have been repaired as well as those that have been marked irreparable and returns with more instruments that need to be repaired.

Instruments to be repaired are brought to a reception warehouse, which is manned by a staff of three. They fill out a form that includes the model number, the serial number of the instrument, the description of the fault, the name of the unit from which it came, and the name and rank of the person in charge of that unit. The form has four copies: two copies are attached to the instrument, one remains in the warehouse, and the fourth is sent to the warehouse for repaired instruments.

Once a day, all the instruments that have arrived that day are moved to the central laboratory for check-out. The test equipment in this laboratory is computerized and extremely sophisticated. Captain Davis, deputy commander of the laboratory, performs the checks and gives the diagnoses. The problems found are noted on the forms. Two clerks initiate a special repair card for each kind of fault discovered and attach the card to the instrument. Instruments found to be functioning properly (i.e., no detectable faults) undergo a second examination. If still no problem is found, Captain Davis decides, according to the description of the fault and the rank of the signatory, whether to declare them to be in full working order or to open repair cards for them. The repair cards state whether the agency responsible for repair is the communication laboratory or the electronic warfare laboratory. An instrument with problems in both communication and the electronic warfare device will carry two repair cards.

About 10 percent of the instruments are declared in full working order and are directly transferred to the warehouse for repaired instruments. When a fault is found only in the communication equipment, meaning the electronic warfare instrument is in good working order, the communication part is repaired while the electronic device goes directly to the warehouse for the repaired instruments. When the problem is only in the electronic warfare instrument, meaning the communicator is operational, the unit is marked for the electronic warfare laboratory to repair and the communication part goes to the warehouse.

The instruments with the repair orders are stored in a warehouse attached to the central laboratory. When the instruments in the warehouse of the central laboratory pile up, the warehouse notifies the staff of the laboratory responsible for repair. The lab responsible then sends a driver to pick up some of the units awaiting repair. Since both laboratories have extremely small working spaces, they usually take a small number of instruments, not more than two working days worth of repairs. There are about 15 models of handheld communicators and about six models of electronic warfare instruments, so the technicians prefer to concentrate on one model at a time. When finished, they start on another model. Centralizing the work that way makes repair personnel much more efficient. That is important because occasionally the tools needed are slightly different, there are different spare parts, and there are totally different technical manuals.

The spare parts are available from a separate warehouse. The technicians withdraw replacement parts on their own authority. They have to sign for those parts. When a spare part is used in an instrument, it is noted on the form. That acts as a counter signature for the spare parts warehouse. Otherwise, the technician has to return the part to the warehouse and receive confirmation.

When work is completed, all instruments—both communication and electronic warfare—are transferred to the warehouse for repaired instruments, where they are stored in an area designated Before Last Inspection. In this area, the technicians attach the electronic warfare device to the communication instrument. Then the instrument goes into the central laboratory for final inspection by Captain Davis.

The technicians declare about 30 percent of the instruments impossible to fix. Most are unable to be repaired because of the lack of spare parts. After approval in writing by Captain Davis, those instruments that cannot be repaired are sent to the warehouse of repaired instruments to be returned to the unit. Davis spot-checks a sample of those instruments to determine whether they really are impossible to repair. He has already "caught" technicians who in the past have tried to make life easy for themselves by falsely claiming that an instrument cannot be repaired. In the past, the usual penalty for such an offense was confinement to camp for two weeks. When the number of such cases increased, the punishment was intensified. Today, a technician caught committing such an offense stands trial and is sentenced to 7 to 14 days in jail if convicted.

Captain Davis offered the following comments about the laboratory, its repair activities, and its problems:

"Since Major Williams handles the new equipment, I have to take care of everything that happens in the laboratory. Major Williams took the best technicians from communication as well as from electronic warfare to work with him. This created an unbearable situation—120 instruments a week is a huge number. There were times when we could fix an instrument and return it to the unit within three days. Today, I estimate the time as two weeks, depending on the model.

"Problems? There is no end to them. First, the warehouse that is attached to the central laboratory, where I sit, is bursting with instruments. At least once a week I have to call one of the technicians and ask him to take some instruments away. I know they do not have enough space in their lab either, but what can I do?

"I get along, more or less, with the guys in communication, but those from electronic warfare are a pain in the ass. There are two technicians left in the electronic warfare lab: Ross and Matt. Ross feels frustrated because he is not with the guys who are working with the new instruments. Two other technicians, Stan and Don, are with Major Williams because they have been here for a longer period of time. So Ross works slowly. The technicians in the electronic warfare lab sometimes need the equipment I use at the central lab. So here comes Ross, who sits around and waits for me to leave the instrument for 10 minutes. He doesn't try to coordinate it by phone, because he claims that I always say I am busy and using the equipment. So there he sits, right in front of me, until I let him check the three or four electronic warfare devices that he has brought with him. What am I supposed to do? I have a lot of work.

"I am lucky because only half the work is done in the warfare laboratory. There are many more faults in the communication equipment lab. Four technicians work in that lab. There is a big mess in the warehouse for the repaired instruments. The instruments must be equipped with their electronic warfare devices. The guys at the warehouse are assholes and don't know, or don't want to know, where to look for the devices. They are usually still at the lab. When the technicians at the electronic warfare lab receive a phone call from the warehouse asking about the devices, they can't answer. They say they are busy. They are full of hot air. The truth is that there are lots of phone calls from the units, and it drives everyone crazy. Matt, the other electronic warfare technician, had an altercation with a colonel who called and demanded to know when his instrument would be ready. Matt got a little fresh, and the colonel wanted to file a complaint. I had to cover for Matt, and I said it was some driver who had come in for a moment and I didn't know who he was.

"Matt is a good, obedient soldier, although he is not very good with his hands. But these are crazy times, and they are driving us all crazy. Yesterday I counted the instruments at the laboratory—there were close to 500! The warehouse is bursting at the seams. I don't know what I will do if I get more instruments. I will have to invade the new building for storage purposes. Major Williams will probably try to kill me, but I would like to see how he would handle this problem."

A Laboratory for the Repair of Communication Equipment—An Analysis

This is a typical realistic case of a service organization experiencing a huge temporary peak of demand. It is intentionally full of details. The idea is to show that TOC provides a way to quickly identify what really needs to be improved. The ability to get a very fast diagnosis is of paramount importance to all of us. It is certainly true for consultants who visit a company and need to diagnose its real problems. The same goes for newly appointed executives who have come from another organization. But, even people who have worked for the organization for quite some time need a different outlook to evaluate the situation of a peak in the load that has changed the rules of the game. In times of change, we need guidelines to quickly identify the core problem within the huge amount of data.

This laboratory has to be treated as a nonprofit organization. The objective of the organization is to repair all the broken communication sets in a short period of time. As part of a global organization, the objective is to assist in achieving the global goal by fixing communication sets. In normal times, that lab is a nonconstraint, maintaining enough excess capacity to support the communication lab's technical needs.

Does the Laboratory Support the Current Demand?

We start with this question because the goal of the laboratory is pretty clear. Now we have to establish what can be done in order to do more for the global organization. It is the global organization that places the demand on the lab. So the next leading question is the one shown above.

If the laboratory is able to meet demand, it means that on average, the input flux equals the output flux. If the average number of sets coming in is larger than the average number of sets going out, the laboratory does not support all the current demands. If the number of sets going out is larger than the number of sets coming in, it can mean only that in the previous periods, more sets were coming in than out and stuck somewhere inside the lab.

All that is known is that the incoming flow is about 120 sets a week. We do not have numbers associated with how many sets are shipped back, including those that cannot be repaired. Can we deduce whether the laboratory supports all the current demand if we do not have this crucial piece of information?

Suppose the input flux is larger than the output. What should the impact be on the work in process? It should grow! We also know that whenever the work in process grows it has to be caused by more sets coming in than going out. We also know that when the work in process increases, the average lead time gets longer. That is a direct relationship. Now we can look for these effects: Do we see any evidence that either the lead time is getting longer or that the work in process is getting larger?

Two statements seem relevant:

"There were times when we could fix an instrument ... within three days. Today, I estimate the time as two weeks."

"Yesterday I counted the instruments at the laboratory—close to 500!"

First of all, it is evident that the lead time is now considerably longer than before. But it is not yet evident that it will get longer. But that can be understood from the second statement. How many weeks of input are stuck in the lab? Five hundred sets divided by the average weekly input of 120 is more than four weeks. Why does Captain Davis estimate the response time to be only two weeks? A reasonable explanation is that when the lead time gets longer, the intuitive assessment is somewhat delayed. Those arguments are mapped in a cause-and-effect diagram (Figure 4.1).

Notice that from a cause-and-effect point of view it is the long lead time together with the input rate that causes the current work in the process level even though we have deducted the length of the lead time from knowing the two other effects.

The important conclusion is located at the bottom: The lab cannot support the current demand.

Is There a Capacity-Constraint Resource?

We have validated that the laboratory is not capable of handling the 120 sets that come in every week. TOC now says that there are two possible causes for that: One of the resources is a capacity constraint, actually a real bottleneck, or there is a policy that prevents the lab from handling more sets.

Let us first inquire whether a policy constraint, rather than a bottleneck, limits the output of the lab. Figure 4.2 outlines my arguments.

What is clear from these arguments is that if we evaluate the situation from Major Williams' point of view, then there *might* be a case to suspect

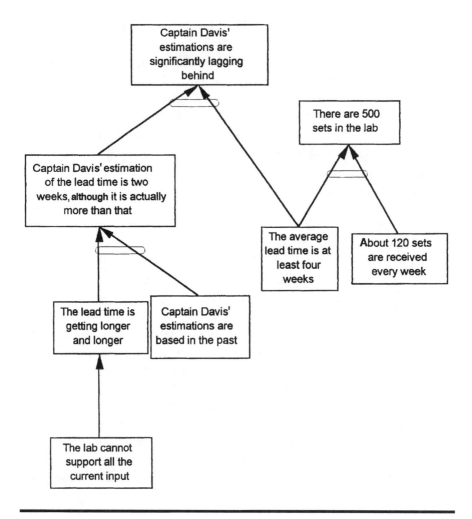

Figure 4.1   Some Basic Cause-and-Effect Connections

that a policy constraint has caused too many technicians to be assigned to preparing for the new generation. Of course, we cannot make a judgment before checking the whole set of requirements and the total available capacity.

Let us first focus on Captain Davis. From that point of view, there is no doubt that there is a bottleneck somewhere within the lab.

How can we identify a bottleneck without measuring the available capacity and comparing it to the load? All we have is a qualitative description of what is going on in the lab. Does the description include effects that can be attributed to the hidden bottleneck?

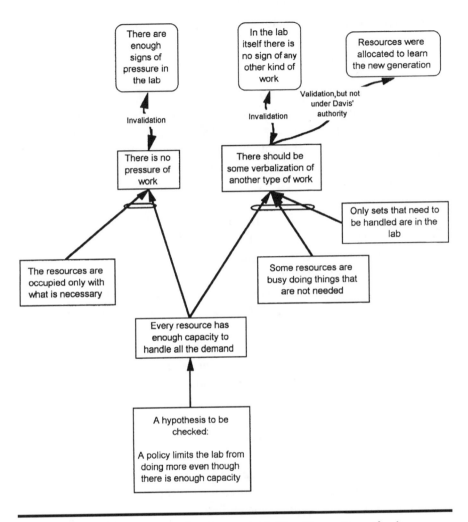

Figure 4.2   Is There a Capacity Constraint? Checking a Hypothesis

What Fingerprints Does a Bottleneck Leave?

Here are some speculations about the possible fingerprints of a bottleneck:

1. A lot of work in process should be waiting for the bottleneck. The physical location of the work in process may be somewhere else, but it is waiting for that bottleneck.
2. When a specific set is searched for, there is a high probability that it is stuck in front of the bottleneck.

3. When we have a human bottleneck, that person(s) feels distress. That distress is radiated in a variety of ways.
4. The bottleneck will be perceived as slow or inefficient.

The distress of Captain Davis and the number of jobs he is personally responsible for makes him an obvious candidate. But careful judgment of the evidence rules him out. Captain Davis' distress is caused by the burden of responsibility, not by the amount of work he cannot complete. Yes, he does say, "I have a lot of work," but if Captain Davis finds time to spot-check instruments that cannot be repaired, then he does have some time to do things that are not absolutely necessary.

But the most crucial information is the location of the work in process. Davis says, "First, the warehouse that is attached to the central laboratory, where I sit, is bursting with instruments." When you look for the signs, you should ask yourself the following: Is the warehouse before the central laboratory or after it? When you check the process, that warehouse contains sets that have already been examined by Davis and are now waiting to be repaired by the communication lab and/or the electronic warfare lab.

If Captain Davis is a bottleneck, that particular warehouse would not be full. Instead, there would be stock in front of the central laboratory, both before first check out and final examination.

So Captain Davis is decidedly *not* a bottleneck. Who is? The work in process locations points to the two repairing laboratories.

Out of the two repairing laboratories, which one produces the far greater trouble?

Here is some evidence from Captain Davis:

"I get along, more or less, with the guys in communication but those from electronic warfare are a pain in the ass."

"The instruments must be equipped with their electronic warfare devices. The guys at the warehouse are assholes and don't know, or don't want to know, where to look for the devices. They are usually still at the lab."

Which lab does Captain Davis refer to? He mentions that the instruments must be equipped with their electronic devices. It comes naturally to him that what is missing is the electronic device, not the communication set. The electronic devices are still in the lab, which is the true bottleneck.

"So Ross works slowly."

"Matt got a little fresh, and the colonel wanted to file a complaint… Matt is a good, obedient soldier, although he is not very good with his hands."

These two descriptions are also caused by the lab being a bottleneck, which makes it look slow.

It can be concluded that the electronic warfare lab is a true bottleneck within the responsibility of Captain Davis. The **exploitation** and **subordination** steps are necessary for any improvement. Here are some guidelines to make improvements:

- The various models of the electronic warfare should be evaluated according to the expected load on the lab. Priorities should be implemented accordingly.
- The supply of spare parts should be subordinated to the technicians of the electronic warfare. That means that whenever requested, the parts will be brought to the lab by the warehouse people instead of having the technicians go to the warehouse to withdraw the parts they need.
- Captain Davis should subordinate himself and the sophisticated equipment to Ross and not cause Ross to waste time.
- If there are sets that need to be fixed for both communication and electronic devices, the communication part should be handled first. If that part is not able to be repaired, then there is no need to repair the electronic device.

When we come to Major Williams the conclusions are now much more concrete. The question is whether the electronic warfare technicians who are with Williams should share some of the load of the repair lab. This is a typical exploitation decision to assess the damage of slow repair of the communication sets vs. slower preparations for the new generation sets.

The policy that sets that cannot be repaired are returned to their units may be viewed as a possible policy constraint. Those sets could supply spare parts for other sets that cannot be repaired. That is a policy that is probably outside the influence of Captain Davis or even Major Williams. But, is it a true constraint? When there is a real bottleneck in the laboratory, it is not clear whether overriding that policy helps to increase the throughput or not. One of the major questions is: Who is going to disassemble the sets that cannot be repaired and test the spare parts? If the technicians who do that are the bottleneck, then they need to evaluate how many sets can be fixed and what the invested time for it is.

This story generates two main messages. One is that real bottlenecks may emerge in surprising locations and environments. The story also demonstrates the devastating effects of bottlenecks in a service organization. The second message is to look for the signs that are caused by an internal constraint. When you know what to look for, there is no need to collect a huge number of facts and data. Any true constraint has a major impact on the organization. It has to be manifest in more than one way.

5 We Need That Department

I am intrigued by hospitals. Not only do these organizations save lives, but they are also quite complex to manage. The next story does not go too deep into the daily operations of a hospital, a worthy subject in itself for TOC analysis. Instead it presents a strategic dilemma. How should a strategy of a hospital be determined in the first place? Moral issues are raised in an organization such as this more often than in regular "profit making" ones. The financial aspects are no less evident, and the conflict between money and medical ethics is very common. The question "What is the value of life?" is avoided, even though it significantly impacts many decisions. And with all these dilemmas around, there is a pressing need for a certain hospital manager to make up his mind.

We Need That Department—Case

The white hair of Dr. Felix Burr is testimony to all the tension he has been suffering since he was appointed Managing Director of Peaceful Hospital in the city of Marjorja. That particular city hospital serves a community of about 300,000 people who live up to 100 miles from Marjorja.

What makes it so difficult to manage a hospital here is the fact that Marjorja is a city of contrasts and with a lot of politics. Some of the notable citizens are extremely rich people who like the beautiful scenery of the nearby mountain and the river on the other side. When the newly equipped airport was constructed in the early 1970s, it signaled a new era in Marjorja life, making it possible for the rich people of the noisy capital to come to the tranquillity and peaceful atmosphere of Marjorja.

This movement of well-established people coming to this area had a dramatic impact on Peaceful Hospital. The new community demanded "adequate" medical treatment, and they were ready to pay for it. So some very good physicians came to Marjorja. Most of them chose to work at the hospital in addition to working for a more profitable private practice. By doing so, they felt they were making an important contribution to the rather poor majority of the town while keeping themselves in very good economic shape. The reputation of the hospital rose, and many young doctors looked to complete their residency program at Peaceful Hospital.

On his 60th birthday, Dr. Burr was offered the post of managing director. He was personally approached by the mayor. Felix was one of the few physicians who had been working at Peaceful Hospital since the 1960s and had kept his position through the change. Felix was a good surgeon and very much liked by his patients and his colleagues alike. Considering the great tension among the departments, the respected and cool personality of Felix seemed to be the obvious solution.

Five years after his nomination, Dr. Burr sat in his office and looked at the agenda for the following day's board meeting. The toughest task was to establish his own mind on the issues and then to control the meeting and back up his view without causing a dreadful argument.

Among the most debated topics for the meeting was to open a liver transplant department. The budget for just the equipment was over $5 million, and it could be obtained by a special donation. Felix knew all there was to know about donations. In that particular case, a special pledge was given to donate $450,000 every year for the next seven years to maintain the new department. Dr. Mark Holden was the initiator of the liver transplant department. Not only he did succeed in finding the potential financial donor, but he also got tentative agreement from the legendary Dr. Bernard Miklehouse to manage the department. For the young Dr. Holden, this was an enormous opportunity to gain international recognition. The topic was heating up the already hot atmosphere at the hospital. Most departments were feeling that such a prestigious new department would get both the resources and the public recognition at their expense.

Felix wondered what his position should be and how he could direct a rational debate on such an emotional topic. Was not this something that was basically wrong? Such a topic needed to be decided rationally. But people treated it as if the donation for the new department would be deducted from their own budgets. And every head of a department thought his department was the most important one in the entire hospital.

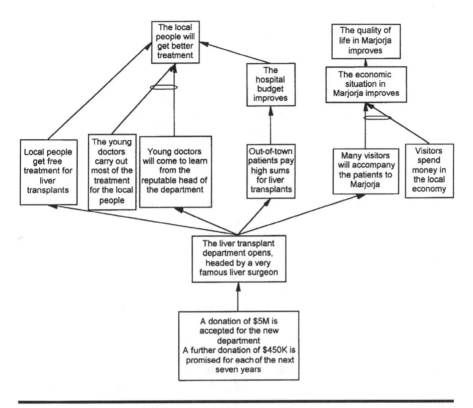

Figure 5.1   Future-Reality Tree—The Benefits

even the remote chance that one of the Marjorja citizens will need a liver transplant, might causes the loss of other lines of throughput.

So the crucial question is: Does the new department reduce the quality of life of certain Marjorja citizens?

If the answer is negative, then *no* conflict should be perceived. The fact that Dr. Burr is not enthusiastic about the new department has to be either because of his concern that some loss will occur or because of personal jealousy. As the managing director of the hospital, Dr. Burr no longer represents the local professional department he used to be with. Let us rule out any personal jealousy Dr. Burr may feel toward such a well-known figure as Dr. Miklehouse. Suppose he really feels that the new department would reduce the quality of life of certain Marjorja citizens. How can the new department damage the quality of life?

Another fact that calls for an explanation is the opposition of the other departments to the liver transplant department. Is it wholly based on personal jealousy? Is the reputation of Dr. Miklehouse so threatening to all the other departments that they object to it so openly? It might also be that the other departments fear that their standing within the hospital will be diminished due to the high prestige of the liver transplant department.

If this were the case, wouldn't you predict that such intelligent people would welcome the new department officially but try to prevent it with less overt activity? It is not politically correct to express clear opposition to something that is perceived as really good and let everybody know the opposition is based on mere jealousy.

One reason for their objection may be that the new department will directly hurt their own patients. That can be explained if there is an internal constraint within the hospital. Such a constraint might also be a cause for Dr. Burr's concern. As director of the hospital, he is in conflict because it is not easy to assess the additional throughput vs. the loss.

Do We Know What the Constraint of the Hospital Is?

There are two possibilities: space for beds and the capacity of the operating theaters. Regarding the space constraint, it does not impact the new department. It can be safely assumed that with the $5 million for the building, the department will have enough beds for the department's patients. So the additional department will not worsen the state of the lack of sufficient beds for the local people. Nor will the new department take highly paid beds from the other departments.

There are enough signs that the operating theater capacity, the other possible constraint, is already a problem. Some operations are done at night. An accident can change the whole schedule, and the need for Dr. Burr to settle the priorities and come up with a new reasonable schedule is the result of a conflict of interests. The lack of theater time is a crucial factor. This effect can also explain the tension in the hospital. Certainly it explains the objection to the new department from the other departments. Liver transplants are known for being long surgeries. So for the other departments, the liver transplant unit is a new competitor for a fully utilized resource.

And Dr. Holden has completely missed Dr. Burr's point. The young doctors who treat to the poor citizens of Marjorja are *not* the constraint. There is no mention of even one effect to doubt the quality of the medical treatment as such. For the poor community, it is the space constraint that really matters, as is the waiting time for surgery, which also negatively affects the space constraint as people stay longer in the hospital.

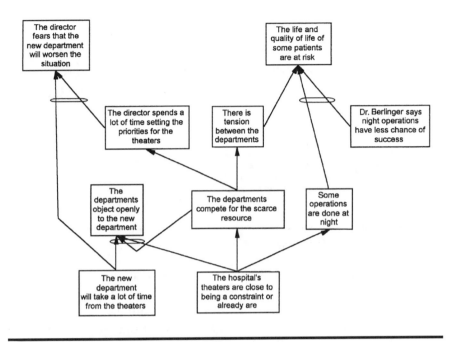

Figure 5.2 How shall the new department worsen an already problematic current reality at the hospital?

So, getting even more bright young doctors is not a real benefit! This is an immediate conclusion from the TOC approach. Dr. Burr knows it intuitively. He is fully aware of the problem of a lack of enough operating theaters. As a matter of fact, this lack is caused by his inability to express the need. As long as all the really necessary surgeries are carried on, it may look good to an outsider. We can assume that the situation is currently not so terrible that people are dying. The theater capacity probably is not a constraint right now, but it is troubling. It is a resource that can easily turn out to be a true bottleneck.

Had the theaters been a true constraint, Dr. Burr would have used this argument already. But he sees a situation where it could reach that level. When the director spends 70 percent of his time on something, you can be sure it is either the system constraint or something very close to it. One can argue that as long as a significant number of operations have to be done at night, with lesser chance of success, then the theater capacity is already a real constraint. But, right now night operations are perceived to be an acceptable solution. If the load is increased, it will clearly become an unacceptable situation for Dr. Burr.

The tree in Figure 5.2 summarizes the previous arguments.

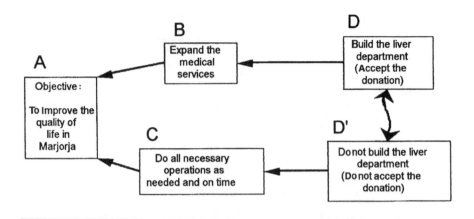

Figure 5.3   Dr. Burr's Dilemma

Taking into account the limited capacity for operations, it is easy to see Dr. Burr's conflict. The new department is an expansion of medical services, and it will bring money, which could be translated into more beds. But the extra load on the theaters will be at the expense of other operations. Can priorities set an optimum operation schedule?

The cloud in Figure 5.3 does not indicate all the assumptions, part of which are stated above. The most significant logical arrow is CD′. The underlining assumption is the emergence of a true bottleneck at the operating theaters and the ramifications that operations will be delayed or done late at night.

Looking for a win-win solution is a characteristic of TOC. This is exactly what injections, or ideas that invalidate certain assumptions behind the conflict, can do. How can we have the benefits of the new department without its negative effects on the constraint of the hospital? We look here for an idea/injection that would solve the CD′ arrow. Not that it is impossible to challenge the other arrows. My intuition would be to go to this particular arrow because I wish to find a way in which the new department will open without all the negative effects.

If a new operating theater is built, it deletes the lower leg completely. That means "evaporating the cloud," or finding a superior solution that transforms the status to a higher level. The hidden assumption this injection challenges is that the current capacity of the theaters cannot be elevated. To really challenge that assumption, we may look for someone who could profit so much from the new department that he will be ready to invest money in a new operation theater.

When such a statement is verbalized, the solution is not that far away. The mayor of the city is clearly interested. Wouldn't it be worthwhile for the municipality to invest an additional $10 million for a new theater at the municipality hospital, and by this gain the added value of the new department to the economics as well as the extra taxes that could be earned from the new businesses? And the shortening of the hospital waiting list for surgeries is worth votes in the elections, too.

Dr. Burr must now set up his proposition carefully: a **conditional** affirmative (yes) to the new department. The condition should be to get an immediate funding for a new operating theater to start functioning at about the same time as the new department. And if the municipality refuses, let Dr. Burr look for other interested parties, including the original donor.

Several valuable messages should come across. The first is to realize how relevant the identity of the constraint is to a dilemma that looks at first like a moral one. Equally important is full comprehension regarding the definition of the goal and the throughput, the added value created by the organization. Then use the conflict resolution diagram (evaporating cloud) to pinpoint the problem and set the direction for a solution.

6 | Where Is My Personal Buffer?

This story concentrates on a part of an organization, specifically the problematic situation of a middle-level manager. The story does not give much information about the whole organization. So the generic questions that were derived at the end of Chapter 1 do not immediately apply here. Still, the story points to a common embarrassing situation that sometimes develops into a real policy constraint. I believe that trying to understand the point of view of the manager of a certain function can give much better understanding to the cause and effect within the global organization. You may be surprised to see crucial generic lessons can be learned from better understanding the perspective of a mid-level manager.

Where Is My Personal Buffer?—Case

As the purchasing manager of All-M, it is my duty to look for the availability and quality of all the materials we buy from vendors. I am also asked to do it at a minimum cost, which constantly puts me in a conflict. Whatever I do, I have to explain my decisions. At any given time, there is either too much or too little inventory, or it is too cheap or too expensive. Dealing with that conflict is the story of my life.

Let me give you a specific story. It may sound more dramatic than necessary because the whole debate was over a mere $50,000. But even that amount of money causes a lot of controversy. Is there a way for a clear and precise decision to be made? It could save me and some other people in finance, operations and marketing a lot of time and probably some health, too.

My story concerns "organic yogurt." As always, it is a combination of our research and development department and the inventive people in marketing. A new product is always a celebration for the company, but it is also a lot of headaches to several responsible employees. This time it was more complex. The organic yogurt was intended to replace our long-time product "healthy yogurt." That meant that once the launch was on, healthy yogurt would no longer be produced.

The preparations for the organic yogurt took almost a year. Throughout the year, I had heard about the advertising campaign. I was mainly involved in the design of the packaging. That was where I was really concerned. All the other materials were common for several products as well as for the two specific products: the healthy yogurt and the organic yogurt. The *real* difference lay in the packaging.

On February 1, there was at last a clear decision to launch the organic yogurt on April 15. The printing design for the new package was not ready at the time. Yet that decision gave me enough time to negotiate the quantity and price of the packaging items with the four vendors that produce them. One of them, PLS-mould, did the plastic cones for our aging healthy yogurt. They have refused to produce the new cones for the organic yogurt because of the complex design the marketing designers came up with.

I was not very sorry about PLS-mould. The company was too old-fashioned and liked to work in huge batches, which did not add to my own health. Instead I've found Agile-Plastic. That company was located some 600 miles from All-M's main facilities, compared to PLS-mould, which is just 20 miles away. But it had a good reputation, so I took the risk of working with them. Organic yogurt was the first contract I was going to sign with them.

I also had to deal with Alum, Inc. That company produced the aluminum lamination to hermetically seal the cones with the yogurt. Fortunately, it had done the old lamination and was going to produce the new ones.

Industry-Print was another vendor involved, and it produced the printing plates for both the plastic cones and the lamination. Viewing the supply chain, you might say Industry-Print should be the supplier of All-M's direct vendors. But, because of the critical importance of the quality of the print, to suit the marketing designer's wishes, I needed to control the quality and flow.

The decision to launch the organic yogurt on April 15 left me with contradicting feelings. If the final design for the new packages had been finished by then, I would have given it a fair chance of meeting the due date. A fair chance from my perspective was about three out of four. The link between the printing materials and the production itself was very fragile, and as many as seven loops might occur. A loop means that in the production, the printing does not prove to be of good enough quality. That happens especially with aluminum lamination. In such a case, the printing materials are shipped back to the printing company for corrections.

Just in time to add more trouble to my mood, Jimmy Coul, the production manager, came to me and said, "Hi, Herb. It won't come to you as a *big* surprise if we aren't ready even by mid-May, would it? We have some problems in stabilizing the production line. The new organic thing is more sensitive to pressure and temperature, and we need to find some engineering solutions for that. I assume that you have your own things to look after. I wish more realistic times could have been quoted, but I gave up fighting marketing for that. Have you noticed how Phil is reluctant to estimate the sales for next year?"

Phil is vice president of marketing. It did not surprise me a bit that he refrained from revealing the forecast for organic yogurt. They had been wrong in the past. Of course, it did put me in an awkward position. When I drew up the agreements with the vendors, there was a lot of meaning to the annual commitment. I could easily get lower prices with a higher commitment. Now it was especially important because the contract with Agile-Plastic had evolved *only* for those materials.

However, my problems with the introduction of organic yogurt were slight in relation to the dilemma concerning the old product! This is why:

All the packaging materials for healthy yogurt will become useless on the day organic yogurt is launched. That means I will have some "dead inventory" that will be worth such and such amount of money to be debited as a waste of money caused directly by my decision.

On the other hand, if on any day there should be a lack of packaging materials, the next day I would be looking for a job. My predecessor had found himself out of All-M for that reason. It did not help him much to claim that a huge fire had broken out at the vendor facilities.

So if organic yogurt was going to be launched on April 15, I would be grateful if all the remaining previous packages went into production on April 14. Of course, I did not believe in miracles. If I were left with just $5,000 worth of useless packages, I would do well. But even that was hard to achieve.

Before the meeting, I looked at my computer to assess how much material for healthy yogurt was already at the warehouse. It seemed we had cones until *about* mid-April. The laminations would be exhausted by the end of March.

The naive reader may suggest something that looks like common sense: Wait until the materials reach the level for the next two to three days, then make an urgent order for a quantity for one day of production. The next day, as long as the launch is not declared, order another one day's supply. Sorry fellows, that doesn't work!

Take the laminations, for instance. They are produced in large cylinders. The minimum quantity is one cylinder, which is the equivalent of approximately a whole month's production. Each such cylinder is worth about $10,000. The printing quality is very demanding, so setting up the machines for the lamination is a long process, and a lot of material is scrapped in the process. Because of that, our agreement with Alum is to order a minimum of six cylinders every time. If I go to Alum with a request for only one cylinder, the company will charge me $15,000! When Abe Markovitch, our comptroller, sees it, he will be all over me.

By the way, Alum's response time is two months. OK, if I press them, they might do it in three weeks.

Now, take the cones. Here, too, there is a minimum order that is equivalent to three months. There is not much point in asking PLS-mould to supply a smaller amount. Since we are going to finish our business with PLS-mould, the company is not going to help me in any way.

You see, even if the date of April 15 is kept, I certainly need to order more laminations for healthy yogurt. I may even need more cones because they may be exhausted a week too early.

Realizing that April 15 is not a certain date at all, what guidelines should I have for the extra orders of packaging materials for the old product? I am especially interested in those guidelines that will keep me in my job.

Where Is My Personal Buffer—An Analysis

There are two troubling questions raised by this case. First, how should Herb make the decisions that are best for All-M? Second, why does management put their middle-level managers in such an unfavorable state? I find the latter question even more troubling.

The first objective of the purchasing manager is to ensure the availability of the materials whenever needed. The management of All-M has taken the precaution of guiding the purchasing manager to the right decision from the point of view of the company—no matter what happens, the materials need to be on time. Herb's complaint is based on the feeling of being unfairly treated because he is measured by his ability to save costs. Herb knows that to fulfill his main mission, *some* costs must be incurred. Hence, the pressure to reduce costs without fully considering the resulting damage is, in so many cases, unfair.

What Prevents Herb From Showing Excellent Cost Control While Ensuring the Availability of the Materials?

There are two main causes for having to spend more money than theoretically necessary. One cause is uncertainty: We do not know how much is needed. The second is that many suppliers make it much more expensive to buy frequent small quantities than infrequent large ones. Hence, there is pressure to buy in large batches. In that case, the uncertainty is much more significant in its effect. Those main causes are depicted in the following cause and effect diagram (Figure 6.1).

The basic cause-and-effect tree simply shows how the purchasing manager is pushed to do actions that also cause a less-than-favorable impression of his performance. We can see that he is pushed to reduce inventories and buy in smaller batches. However, if he does that, he will still be poorly judged. Hence, this tree directs us to draw the main conflicts of the purchasing manager. We show here the conflict that concerns the uncertainty.

The inventory management dilemma is very common. The delicate role of inventory management is to be a protective mechanism against uncertainty. If the market demand and the production planning could be specified very accurately, a straightforward mathematical equation could specify how much to buy and when. As uncertainty is a basic ingredient of our life, inventory management is not simple. The upper part of (Figure 6.2) outlines the need to look for the availability of *all* the necessary items for production. The bottom part

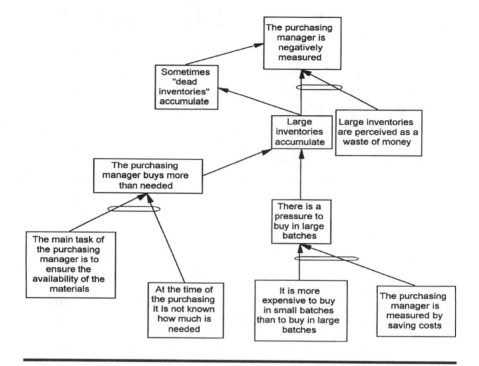

Figure 6.1   Basic Cause-and-Effect for the Purchasing Manager of All-M

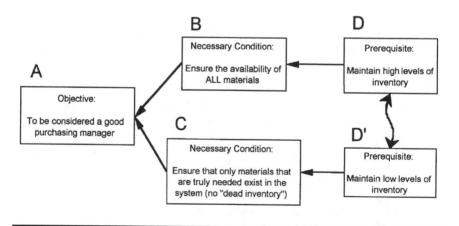

Figure 6.2   The Common Inventory Management Dilemma

relates to the cost of maintaining inventory. The "dead inventory" issue is written because of its significant relevancy to the dilemma. A more generic necessary condition would be to ensure low material costs.

Of the various assumptions behind the arrows, I would like to mention the one behind AC. In the majority of the conflicts, the proposed injections usually challenges either the BD or the CD' arrows, but in this particular dilemma a more profound basic assumption may be rethought. The assumption is that high-level management cannot assess the amount of uncertainty at the lower levels. Another basic management assumption is that it is of *significant* importance to cut costs *everywhere.* Hence, high-level management tends to downgrade the amount of uncertainty in the lower levels and expects that no, or very little, safety stock to reside at the warehouse. Hence, in order to be well-appreciated, the purchasing manager should refrain from adding safety stock, which is the context of C.

The following analysis will be based on the following generic ideas:

1. Uncertainty means that certain variables (for instance the exact launch date) may get different values. When one mentions such a variable, in most cases it is understood to be the predicted expected value, in other words the average.
2. Uncertainty causes a lot of damage.
3. The *actual* value may be less or more than the expected value. There are usually different damages when the actual value is less than the average from the damages when it is more. In most cases, one of them is significantly more damaging than the other.
4. It is possible to implement a protection mechanism against the higher damages of the uncertainty.
5. The protection mechanism costs money. Hence, it should be clear that on average implementing the protection mechanism will save money by reducing the average damage by much more than its cost.
6. Management must be better aware of the need to implement the protection mechanism and must be consciously willing to pay the price.
7. The effectiveness of the protection mechanisms can be monitored and controlled.

Those ideas are the kernel of the idea (in some TOC circles the term injection is used) that should challenge the AC arrow. The purchasing manager's decisions will be described and analyzed in detail to demonstrate the rationale.

Within the context of this story the punishment for lack of material is much more severe than the punishment for having too much inventory. I assume All-M realizes that lack of their yogurt might force their loyal customers to switch to one of the competitors' yogurts—and maybe never come back. Hence, in many cases the damage caused by lack of materials is of greater magnitude than the damage caused by some excess inventory. In guiding the purchasing manager to make sure that all the materials are available, the company's management is doing the right thing. In other aspects of the management behavior, it is the other way around.

Herb's actual complaint is that he is in a position where he is certain to lose. What are the chances of having the exact amount of the older healthy yogurt packing materials? Very close to zero. Given the nature of the purchasing order batches, he is going to have a very significant amount of useless packing materials. The problem is that management ignores those expenses that stem directly from its decision to move from the healthy yogurt to the organic yogurt.

Let us deal with the question of how Herb should make decisions that are good for the global organization. Naturally, we still have a conflict between too little safety and too much safety. The question is, can we come up with a good enough decision regarding the right amount of packaging materials to be purchased?

From what we know, the predictions are that the current laminations stock is going to last until the end of March. It seems that there is a need to order more laminations. The specific questions now are: When is the right time to order more laminations and how many cylinders are needed?

First, let us deal with the timing. Herb says that with the appropriate pressure on Alum, it will deliver the supply in three weeks. But the regular response time is two months. Suppose it is now February 1, should Herb wait to the last minute or order two months in advance, meaning immediately?

The predictions are that the laminations stock will be exhausted by the end of March. The launch of the new yogurt most certainly will not happen before April 15. Is it unreasonable to assume that *maybe* the current inventory will last until then? Right now, the stockroom has laminations for two months *on average*. Could it last for two and a half months? Probably, and maybe even more.

I do not believe that going through the statistical models will really help Herb come up with a reasonable decision. The problem is complicated, acquiring the data takes a lot of time and effort, and the decision itself is not that critical as long as one acts with common sense. TOC teaches us to look for a reasonable way to simplify the situation. Let us look for some reasonable rules for decision making in such an uncertain environment.

In an uncertain situation, there is a clear advantage to delaying the decision. The rational is that the later the decision, the better the assessment of the actual needs. In this particular case, the later Herb orders, the better his assessment will be of how much (if any) he should order. Of course, Herb should not delay it to the extent that all the inventory will be exhausted. So, we need to determine the latest time to order that still provides very high chances to be on time. Alum can be pressured to deliver in three weeks, and in this case any surplus inventory is going to be scrapped. Herb should use that option. However, Herb cannot fully rely on the three-week delivery. That number is probably an estimate of the average time for such an order. Some extra time should be added as a protection mechanism.

Herb's action to issue an order should be initiated by the current stock going below a certain level that is an "emergency level buffer." The inventory level when a new purchasing order is issued is a common and crucial protection mechanism. As long as there is still stock, the company is protected from too high demand or too slow delivery. The determination of that level needs to ensure that the shipment of new laminations will arrive *before* the current stock is fully exhausted. The term "ensure" has to be limited. If it is necessary to ensure 100 percent availability of materials, Herb should have ordered a year in advance. Even then there is some probability of missing some materials. In all our decisions, we cannot fully ensure anything. We need to assess a "reasonable" certainty. What is the level of a reasonable certainty? Is it 90 percent or 99.9 percent? It depends on what is at stake. As human beings, we do have some intuition to judge what is reasonable and what is not. The term "beyond reasonable doubt" is used in the justice system in a similar way here. The jury cannot say whether they are 99 percent or only 98 percent certain of the guilt of the defendant, but they are still able to determine whether the doubt is reasonable or not.

Please note, the above determination is not for an average assessment of how much stock is needed until the next shipment arrives. It is definitely greater than that. It is intentionally made so that in the vast majority of cases, some old stock will be there when the new stock arrives. How much stock? This is the "safety stock" that was needed because of the uncertainty.

The purchasing manager should rely on his intuition to determine the order level for the laminations. That will dictate the latest date possible to order but will still protect the availability of materials.

How Many Cylinders Should Be Purchased?

The crucial factor is the launch date. The order should be issued when the stock is at or below the emergency order level. Even when we are confident that the current stock will last until April 15, we still need to know if the launch date will be delayed.

Again, the question is to assess the *latest* reasonable date for the launch, not its average (expected) date. If the launch is delayed, many more old materials are needed to enable safe continuation of the production of the old product.

Determining the latest reasonable launch date is not enough. The other factor is how many cylinders are to be consumed during this period of time.

The decision is for a combined protection mechanism, which is determined very much the same way as determining the inventory level to initiate a purchasing order. How many cylinders will ensure that the laminations will not be exhausted before the launch? If the laminations are exhausted, more cylinders will have to be ordered and probably at higher prices for less than the minimum number. And the probability of having too many cylinders will go up. To prevent that from happening, the current order should protect the company from the impact of two independent occurrences: a delay in the launch date and an increased market demand.

In this case, the production manager predicts the launch to be about mid-May or even later. Suppose it is not reasonable that the launch will be delayed later than July 1. If it is March 1 and there is about one cylinder left (which covers a month's production on average), we will need three cylinders to cover the three-month time period until July. That does not give reasonable protection because the demand might be greater than average. And if that happens, the damage to the company is huge. Hence, we may need four cylinders.

Should the purchasing manager order four cylinders on March 1? All-M will need to pay extra because the order is less than the minimum required by the supplier. Still, if we are almost certain that we will not need more than four cylinders, it is better to pay more for four as long it is cheaper than five.

If the decision on March 1 is to buy laminations only once, the order should be for four cylinders. The extra cylinder is a protection mechanism that costs money, but it is worth it because it *may* prevent much more significant damage that is difficult to quantify.

Is This the Best Common-Sense Decision from the Global Organization Perspective?

The four cylinders represent the worst reasonable scenario. It certainly may be the case that on April 15 the launch will take place. In that case, about three and a half cylinders of laminations are "dead inventory," nothing can be done with them.

That is the cost of uncertainty, right? It is known *a priori* that *maybe* three and a half cylinders of laminations of the old product will be dumped. But, at the time the decision is made, it could also happen that more than three out of the four cylinders will be used for production. This extra cost is still a good enough management move in the uncertain circumstances.

What Can Be Done to Reduce the Costs of Uncertainty?

In this case, we may consider the possibility of ordering only one or two cylinders, paying much more, and paying even more when the need for more cylinders is established according to the above rule (the emergency order level).

The difference between ordering all four cylinders in March and ordering only two and then another two later depends on the extra costs of ordering small batches from Alum. It highlights two guidelines in making decisions with a sense for uncertainty:

1. Consider both the "reasonable best scenario" and the "reasonable worst scenario" for the decision making. I call it the "reasonable range."
2. As time goes on, assessments may be improved. Hence, the cost of making a decision with a sense for uncertainty can be reduced if the decision, or part of it, can be delayed.

I assume the reader can use the same reasoning for the cones. In that case, the question of whether to order at all is even more critical because it is possible the current stock will last and the magnitude of the order is less flexible because of the relations with PLS-mould.

How Does All This Relate to the Theory of Constraints?

In my opinion, the above arguments are TOC reasoning that go all the way back to the five steps. I interpret the management guiding rule that the lack of materials should not occur no matter what as a **subordination** rule. What it says in TOC terminology is that materials should never be the constraint of All-M. The identity of the system constraint of All-M is less important to the purchasing department. The department needs to subordinate to whatever requirements stem from the exploitation of the constraint.

The requirements that are derived from the system constraint do not give the purchasing staff enough time to buy everything according to a clear demand. Hence, the situation with the purchasing department is that it needs to base its decisions on uncertain predictions. That is the most common state of the purchasing department in organizations. It means that good subordination processes need to be established to ensure the availability of materials while keeping the costs under some control. That is the heart of the conflict depicted in Figure 6.2.

The protection mechanisms are part of the subordination processes. The task of the subordination is to support the exploitation of the constraint. The greatest damages to the organization happen when the exploitation scheme fails. Because uncertainty may be a major cause to the failure of the exploitation scheme, the scheme needs to be properly protected.

TOC addresses that problem in certain areas. The concept of the "buffer" is of great importance in the drum-buffer-rope, the TOC-based planning methodology for the shop floor, and in the TOC project management methodology. The way the concept of the buffer is used within the shop floor is expressed in time units and is an estimation of a fairly long lead time from the raw materials to the protected area. The buffer represents a reasonable worst case of how much time it takes for the materials to get to the constraint or the shipping area. This estimation is based on the intuition of the production manager. TOC then uses the buffer time to prevent the materials from being released too soon.

Why Does Management Put its Middle-level People in Situations Where Every Action They Take Seems Faulty?

My hypothesis is that the management does that because it cannot quantify the uncertainty the lower-level manager faces. It believes that pressing the person from both sides (availability and cost) will make the person come up with the best decision.

I don't think so! The above analysis was made under "no fear" conditions. Hence, the possibility of ordering just one cylinder and paying extra money was fully considered. If Herb fears Abe Markovitch, he may order six cylinders and pray for a huge delay in the introduction of the organic yogurt. He might also do well, in his own interest, to order the materials as early as possible. It will reduce his superior's criticism when an order for items that eventually were not used was placed long before the switch in the products. Such behavior goes clearly against the interests of All-M. So to whom do we expect

Herb to give his loyalty: to himself or to the organization that does not handle uncertainty very well?

The sin of too many managers is to ignore the impact of uncertainty. They demand from their people reduced costs to maintain a reasonable protection mechanism. The ignorance concerning uncertainty causes fear. When a capable manager looks for luck to be considered a capable manager, it leads to some devastating effects. The wish for a personal buffer (protection against uncertainty) has an enormous negative impact on organizations.

For instance, Phil, the vice president of marketing, refrains from revealing the forecast. Why? Because the forecast always lies, and he might be perceived as being responsible for actions that were taken because of the forecast. The damage caused by such actions is huge. This is a very bad subordination to the system constraints, which need the best planning possible.

The above analysis has exposed the cost of maintaining a protection mechanism. The actual expenses may increase due to the difference between the best and the worst case scenarios. Once the expenses of maintaining appropriate protection are revealed, the following issue arises:

To provide the management with some control mechanism on their people who make decisions in uncertain situations that threaten the constraint, we should look for a way to assess/quantify uncertainty in situations where statistics cannot be applied.

If an agreed upon way to assess uncertainty is established and implemented, the conflict will be eliminated. The decision making will be clear, and Herb will not have to worry about how he is perceived by his superiors.

The rough idea of maintaining a protection mechanism is based on the intuitive assessment of the reasonable worst case and reasonable best case. Suppose every manager needs to justify his actions by providing such assessments. When the actual result is known, it can be put in relation to the predicted range. Manager assessments can be monitored and over time can be analyzed statistically and may produce valuable feedback on whether the assessments were too cautious or too daring. That is the basic idea behind buffer management, the control mechanism offered by TOC for the drum-buffer-rope and critical chain methods. (See *Re-engineering the Manufacturing System, Applying the Theory of Constraints* by Robert Stein and *Project Management in the Fast Lane: Applying the Theory of Constraints* by Robert Newbold.)

And yet a question may be raised after we acknowledge the cost of maintaining protection mechanisms: What could have been done to reduce the level of uncertainty?

The benefits of reducing the uncertainty not only save the expenses of the protection mechanism, but they also improve the subordination processes. First of all, they reduce the fear within the organization, then they increase the chances for better exploitation of the constraint.

The more flexible the vendors are, the less uncertain the purchasing manager will be. The more significant factor here is time! If Herb could issue an emergency order with a surcharge of 10 percent that would be supplied in 5 days, the problem would almost disappear. Of course, the possibility of ordering just one cylinder is important. Without it, the uncertainty expenses will go up considerably.

So, how much is it worth to companies such as All-M to have very flexible vendors even though flexibility has a price tag to it? To answer the question, the damage of maintaining the protection mechanism for the higher level of uncertainty needs to be evaluated. To do that, one needs to assess the magnitude of the uncertainty. If this is so important, it can be done.

As you have found out, this case was used to bring up the issue of making decisions in uncertain situations. The ideas expressed here have been initiated by the TOC techniques but were not verbalized in this way. What do you think about it? I treat this book as a learning platform. The stories are used to provoke thinking. I am not necessarily right, but the issues are real and important.

7 The Perspective of an Organizational Change

T *his story is a about a change that did not produce the desired expectations. Why? The underlining ideas behind the change are popular and perceived as very beneficial. Can the world be wrong? Or the way the change was implemented was wrong? Do we have the tools to make a sound decision?*

The Perspective of an Organizational Change—Case

An hour before the Sound-Soul board of directors' meeting, Aaron Bikhof, the founder and president of the company, was sitting in his spacious office thinking about the previous two years since he had initiated the organizational changes at the end of 1995.

Aaron was not pleased at all. Of course he did not even think about showing anybody that he had some doubts about the dramatic process that had turned Sound-Soul into a decentralized organization built upon profit and loss centers.

But Aaron wanted to be honest with himself. His parents had given him the sense of value of a person who knew how to criticize himself and learn from the process. You could not learn from your experiences if you could not admit to yourself that you had made a mistake. Life had taught Aaron that you did not have to scream at the top of your voice to announce your mistakes. At the coming meeting, he did not intend to say that he may have made a mistake with the whole action.

Had it Really Been a Mistake?

The simple facts were that the company had shown a reasonable profit in 1996. In 1997, it was expected to show a smaller profit. But from an objective angle, the situation was not that bad.

But two years before, the hopes had been on another scale. Then, Sound-Soul had been very profitable, although some decrease in profitability could be felt. Aaron had approached Eddie Small, a managing consultant whose specialization was both financial and organizational. Together they had fashioned the new move, which was meant to renew the momentum of Sound-Soul. The idea had been to build four main profit centers. The marketing and sales department would function as a marketing and distribution profit center of the stereo products of Sound-Soul. The three factories—the plant for the disc players and tape recorders, the plant for the volume systems (amplifiers, tuners, and receivers), and the plant for the loudspeakers—would be profit centers that would sell their merchandise to the marketing and distribution profit center.

The move brought about a considerable delegation of authority to the four managers of the centers. Aaron, who up until then had been making most of the decisions himself, had to tell himself that the real challenge was now not to interfere, unless...

The most difficult point during the process was deciding on the transfer prices on transactions among the different bodies. Sound-Soul sold mainly combined minisystems made up of a compact disc player, a tape recorder, an amplifier, a tuner, and two loudspeakers. The target customers of the company were especially the youth. The company's products were accordingly cheap and were directed at a relatively small number of models sold in large quantities. The combined product and the functioning of the marketing and distribution department as an independent center necessitated a high level of integration in spite of the fact that each one of them dealt with a different technology.

In the economic research he carried out, Eddie found out that the disc player plant was on the borderline. It seemed that in that area, the customers did not wish to spend more for a higher level of equipment. The tape recorders were not very popular either, and Sound-Soul was already selling many systems without tape recorders at all.

The sound system plant was relatively profitable, and the loudspeaker plant was clearly profitable. The calculation of the product cost of each item was based on a careful examination of all the functions needed to produce it. That examination took into account the development and research functions and their contribution to the products. The original account stated transfer prices based on the full cost of a single product, as calculated by the activity-based costing procedures, with the addition of 15 percent profit. Those prices created a problem: The price the marketing profit center had to pay was substantially higher than the market price for the compact disc players.

Since Aaron was determined to keep the disc plant, a compromise was achieved: only 10 percent profit for all the factories. The loudspeaker plant suffered most. The body that gained more than anyone else was the marketing profit center, which was the one with the higher profit potential. Aaron decided to apply the compromise proposal on an experimental basis and to examine the situation again in two years. The moment of the examination had arrived, the moment of truth, and Aaron was wondering what the correct solution for Sound-Soul was.

The main line of products of Sound-Soul was the one of the minisystems. There were three different models, and the prices were between $250 and $600. The minisystem line was 60 percent of the company's sales.

A line for midi-systems consisted of more expensive systems and formed 20 percent of the sales.

During the two previous years, after the organizational change to profit centers, the loudspeaker plant had offered the public a new product: two big loudspeakers sold separately without the combined systems. That introduction had attained great success: 12 percent of the company's sales today were from those loudspeakers.

An additional 8 percent of the sales were sales of parts of the systems: disc players, receivers and loudspeakers.

The loudspeaker plant, which had presented the new loudspeakers in April 1996, had increased its profits considerably. The transfer price of the loudspeakers was decided according to the relation between final sale price and transfer price of the regular speakers when sold as parts. Both the marketing and loudspeaker plant claimed that the decision deprived them of profits.

The disc player plant showed a loss in 1996, and it was expected to have higher losses in 1997. The receiver plant was on the verge of breaking even. The marketing profit center had kept its profit level.

The entire company had lost some of its profitability during 1996, and this trend was supposed to continue in 1997. There would still be a profit in 1997, but the mini line was in a slow but constant decline. There were more sales of parts of systems. It was obvious that the success of the line of the big loudspeakers had slowed down the decline. But the managers of the disc player plant and receiver plant claimed that this success was the reason for the almost-failure of the mini line.

The meeting was called for September 1, 1997 to sort out all the arguments and claims and also to discuss two new suggestions: the manufacturing of loudspeakers for Crox, a company that directly competes with Sound-Soul, and a rapid development of a new minisystem, which hopefully would help the company gain back the market segment it had lost. Before the meeting, Aaron was wondering what he should do. It was easy to guess what the plant managers would say, they had clear interests. The question was, Aaron said to himself, "What is good for Sound-Soul?"

The following are parts of the minutes of the meeting:

Alfred, manager of the disc player plant: "I feel I have to apologize because my plant is losing money. But the calculation we made back in 1989 shows that it pays more to produce whole systems than separate parts. I know this is not what they think in the market, but I must remind you that we sell relatively cheap systems. In our area, if you buy the components for a system in a shop, you pay about 8 percent less than for an integrated system. So why do people buy integrated systems? Because the customer does not know how to choose the components of the system and also because the system planned as a complete system looks more compact and aesthetic than a player, receiver, and loudspeakers, each from a different company. Notice that the market for relatively cheap parts is also very small. In the cheap segment of the stereo systems, you can mainly find whole systems. All our competitors sell whole systems, even those who manufacture only part of the system purchase the other parts and integrate them into a combined system.

"According to this analysis, what is happening to us is causing damage to the market and to our functioning. The sales of the minisystems are decreasing because the loudspeakers do not arrive on time. I suggest we pay attention to the drastic decline in keeping the supply dates. Since the integration of the system is made in my plant, I am in the position to tell you that we have a lot of completed disc players and ready receivers, but we do not have completed integrated systems. What are we waiting for? For the loudspeakers. Since they have started to occupy themselves with the big ones, it takes all their testing time. The quality inspection of the big loudspeakers takes a lot of time, so the small loudspeakers and the systems pay the price.

"I know that everybody thinks that the price I get for the transfer is relatively high, well you should know I can get this price outside if I commit myself to very large supplies. Today I have 20 percent excess capacity. Any contract I can get outside, I have to commit myself to at least 40 percent capacity. I am not prepared to screw Sound-Soul. I also want to say that if somebody believes he can get disc players outside for a lower price, he is mistaken. What you can get is manufacturing remnants at special prices. What you cannot get is a clear and loud commitment for quantities and to keep the supply date. I also wonder who will be ready to manufacture disc players in yellow and orange. Have we forgotten that great part of our success has been due to the young design of the systems?"

Roman, manager of the loudspeaker plant: "When you succeed, there is always a lot of jealousy. Alfred complains about the availability of the loudspeakers. But we worked exactly according to marketing's forecast. If the market shifted from C4 systems to C5 systems, I would have expected Simon to know it beforehand.

"True, we are working very efficiently now. In spite of the ridiculous transfer price, we show a profit—and this profit goes straight to Sound-Soul.

"When you work efficiently, and we keep the manufacturing lean, there are sometimes delays. But the bottom line is we are the ones who contribute more to the company's profit.

"Now, there is Crox's offer. We are talking about loudspeakers similar to our C5 systems. Larger quantities in production will ensure more efficiency and also a running supply for Alfred. All we have to do is to work a third shift in three working stations. The financial report was approved by Eddie Small, so this move looks very profitable. Note that for each C5 loudspeaker, I get $72. Crox pays $76, and this is significant. It adds $4 to Sound-Soul's profit."

Simon, manager of the marketing and sales department: "If the loudspeaker plant sells to Crox, it will damage the sales of the mini lines, since Crox sells to the same public we do. If we get $76 for loudspeakers for a system Crox sells in place of us, we have lost a sale of $380. Is this worthwhile?

"The large loudspeakers, the G line, are selling fairly well. But even before the financial analysis, they are only slightly more profitable than the C5 system, and C6 is even more profitable. I do not know if the product's life is for the short or the long term. The G line entered the market with good timing; there is a demand for big and cheap loudspeakers. But there are other companies that specialize in loudspeakers, and I am afraid they will declare a price war on us. If, meanwhile, we have destroyed our mini line, that victory could destroy us.

"As far as marketing is concerned, the mini line has started to drop. Correct, supply problems have greatly contributed to this, but the line has also started to decrease in its life cycle. The tendency toward C5 means going in the direction the public wants. We need something better designed in order to get a higher price for it. We must cross this line. Next year we will have to come up with something."

Sharon, financial manager: "I want to clarify the profitability issue. According to the calculations of the activity-based costing, loudspeakers of the G line have some advantage over the C5 systems. The G line today is approaching the margin of 18.2 percent. The C4 systems have margins of a mere 1.2 percent, C5 systems of 15.6 percent, and C6 systems reach 20.3 percent.

"But those calculations do not show the whole picture. We have excess capacity in both plants, and this does not give any additional value to Sound-Soul. We have already discussed that it was advisable to leave the C4 systems but only when we have something to sell in place of it. If we could sell more C6 systems, we could make much more profit. Maybe Simon should be aware of this fact. The real star is the C6 systems. When the loudspeaker plant works at full volume, it is good. But that does not mean that closing up the disc player and the receiver plants will increase our profits. In that case, the loudspeaker plant would have to find many more customers than it has today. I am not sure Crox will transfer all of its loudspeaker line to Roman. Moreover, it will then be in a position where it could push for a very low price. I am not certain about the future of the loudspeaker plant working on its own."

Aaron, company manager: "The large loudspeaker models of the loudspeaker line are a very positive phenomenon. It is to make possible these kinds of initiative that I went for the organizational change. The large loudspeakers' profitability, which surpasses our other popular products, shows that this has been a very smart move. I expect much more activity in the disc player and in the receiver plants.

"What bothers me are the petty arguments about the transfer prices. We have already said that this system for transfer prices was not exact and that we wanted a way that would possibly compare with the previous year. If Alfred's plant is losing this year, this is not right. There are all sorts of excuses. I do not mind if Alfred purchases loudspeakers from Crox. I still want to think about the subject of Simon selling Crox loudspeakers. There are considerations for and against this.

"I want cooperation among the plants. All these arguments interfere with this. To me, it is clear that a new line of systems should be started. This forces all the profit centers to cooperate with one another. We do not have enough human resources for the development, and we also need new resources, meaning a development department. That is a considerable investment, and

I believe the four centers should participate in the investment as well as in the development efforts. If you are not able to reach an agreement, I will decide on the amount of investment from each of you. I am convinced that we must invest in development. And if you do not reach this conclusion by yourselves, I will again have to become the dictator of Sound-Soul. I simply cannot understand it."

The Perspective of an Organizational Change—An Analysis

You already know the regular question for every organization's manager: "What blocks ...?" It is certainly relevant for Aaron in this case as well. But this organization experienced a considerable organizational change in the past two years. Instead of starting from scratch, we should try to first evaluate how the change impacted the company's overall performance? This is the question Aaron is asking.

Was Reorganization Into Profit Centers a Mistake?

What do we know about the outcomes?

- The global profitability of Sound-Soul is going down.
- The market for cheap minisystems is going down.
- There are ongoing arguments among the centers regarding the transfer prices.
- There are on-time delivery problems.
- There are many disc players and receivers at the integration step—waiting for loudspeakers.
- There is no agreement among the managers of the profit centers regarding mutual development of new integrated systems.

There are more facts, of course, but these suffice for now. Looking at that list of undesirable effects, could they be caused by the change to profit centers? Figure 7.1 shows a part of the current-reality tree. Although it encompasses all the undesired effects on the list, it does not address why only the loudspeaker plant offered stand-alone products. It merely states that the disc player plant and the receiver plants do not have a good market for their components. That claim is supported by Alfred's analysis, which I chose not to include in the tree to keep it direct and simple.

The most significant undesirable effect is that the global profitability is going down, which is at the top of the tree. It is assumed that the combination of market demand, which is decreasing, and the reluctance of the managers to cooperate on the development of new integrated systems is the most significant cause for the profitability drop. One could argue that more aggressive marketing, especially of the more expensive sets, could generate more profits. One can also challenge the assumption that the market for cheap, separate disc players and receivers is low. Thus, if every plant introduces its own stand-alone products, the company will show more profit.

This book is about TOC thinking guidelines, not about the characteristics of any specific market. Let us accept the basic assumption, which Aaron shares, that in order to fully succeed in that market, integrated systems are crucial. That does not mean that no stand-alone products are to be introduced, but that the main business line should be integrated systems.

What is the most significant root cause to handle? To impact the whole market of cheap minisystems seems too ambitious. The root cause—"Every profit center is concerned only with its own performance"—causes the whole list of undesirable effects.

What is the rationale behind the above claim? The profit center is declared as an independent organization. The manager of a profit center gets much more authority to act within the specified set of rules. The profit center is requested to prove its economic desirability. It is measured as if it is an independent organization. So, profit centers are concerned almost solely with their own local interests.

If Figure 7.1 represents a valid scenario of what happens when an organization is split into several profit centers, why does it seem like such a good idea? Here is the conflict that Aaron finds himself facing:

Figure 7.2 depicts a fundamental conflict that is also in another case. The generic conflict is between individual achievement and team achievement. The B part of the conflict claims that the CEO needs to ensure the individual motivation of the local managers to excel. The term "local" is used here as an entity that is a part of the larger system, not necessarily in the geographical sense. The most basic assumption here is that there is a need for local managers who are capable of making intelligent decisions. Another important assumption behind the AB arrow is that the local managers need to have the right motivation to do their best. That means that such high-level motivation is not taken for granted. It is also assumed that local managers need to be at their best to help the global organization excel. TOC has some reservations about that assumption. TOC distinguishes between the constraints and the nonconstraints, claiming that only limited benefits stem from improving the nonconstraints. However, the subordination needs are such that the organization requires good enough local management to be able to properly subordinate to the constraints (including the market constraint). This in itself demands a good and highly motivated manager.

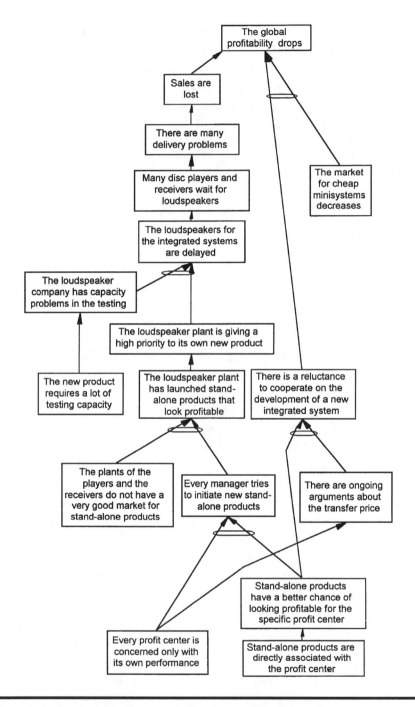

Figure 7.1   A Concise Current-Reality Tree of Sound-Soul

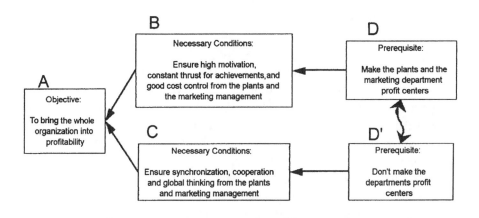

Figure 7.2 Aaron's Basic Conflict

From that basic need for creating the high-level motivation to excel, another assumption is generated: The local managers are appropriately motivated based on the authority they are given! The higher the authority, the more motivation we can draw out. Giving much authority to local managers is not enough to ensure motivation in the right direction; there is a need to measure their success. Two reasons for that: First, we need to have adequate control whether the manager is capable or not. Second, without clear success measures, there might be no motivation in spite of the given authority. Figure 7.3 draws the assumptions behind the ABD arrows.

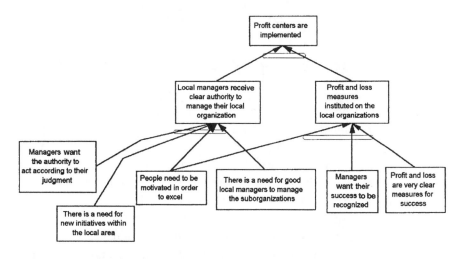

Figure 7.3 Why Profit Centers?

The bottom part of the conflict (Figure 7.2) contains another set of assumptions. Here the need for synchronization between the suborganizations is emphasized. The most important assumptions exist behind the CD' arrow.

What is behind the perception that profit centers interfere with the synchronization of cooperation throughout the company? Because every profit center is measured independently from the others, every profit center manager looks hard at his own profits. That brings us to the root cause of Figure 7.1.

A highly debatable assumption behind the BD arrow, which also appears in Figure 7.3, is that "profit and loss are very clear measures of success." To measure the profits generated by a department, all one needs to do is create a situation in which such a department is an independent organization that needs to find customers to pay for the added value it can supply.

Is the profit and loss statement a good measure for part of a larger organization? Is "selling" a product/service within the global organization the same as selling in the free market? In the free market, the prices are determined by the interaction of buyers and sellers. That is not easy to imitate within an organization unless the profit centers get absolute freedom to sell and buy without any preferences to the other members of the organization. Another way is to base the transfer price on the product cost, plus certain profit margin. But TOC claims that there is no valid individual "product cost," the rationale of which is described in Chapter 1.

The assumption that it is possible to assign a "fair transfer price" for every product of the profit center is necessary for maintaining a good measurement system, via the profit and loss statement of the individual center, to motivate the local managers to do their best for the global organization.

Is It Possible to Come Up With the Right Transfer Price?

The tricky word here is "right." From what perspective is a transfer price right? The most obvious meaning is that such players can be bought from other vendors at the same price. As Alfred, the manager of the player plant, noted, one has to check whether all the company's requirements, quantities, quality, and lead time, are satisfied. Suppose that the transfer price for the players is right in view of all the requirements, but the plant still suffers from losses. Does it mean the plant should be closed?

What do we lose when we buy components from others? Some managerial control, for instance. If there is a need for a very urgent delivery, it is up to the good will of the supplier to respond positively, and it might cost extra.

What if our supplier finds a much better client and the reliability drops. Is it easy to find another supplier under the same terms?

And what about developing a new integrated system? If the four profit centers had been four independent companies, I guess it would have been up to the marketing company to dictate the global design and look for the synchronization among the three plants. Only rarely is there full synchronization between independent companies on the design of a new system.

And there is the possible loss of business opportunities, as well. Suppose every profit center gets its right transfer price. Suppose marketing finds an opportunity to sell systems for a segmented market niche, but at 20 percent price reduction. Will all the plants agree to accept the reduction on their transfer price? If they do, won't marketing start to use it extensively, thus reducing the total transfer payments? This will lead the plants to a loss while marketing makes a nice profit. If they refuse to cut their transfer prices for the special opportunities, marketing may realize it might show a loss of the deal and reject it. Is such a deal worthwhile for the global company? It depends on the circumstances, which means it is sometimes worthwhile (especially when regular sales are down and there is excess capacity everywhere).

Let's now deal with the question of how to determine the right transfer price. In the free market, prices are determined by negotiations. The full product is not just the player or the receiver, but the commitment to quality, quantity, and lead time and the trust between the supplier and the client's buyers. It is not simple to find out what agreement you could have reached with a vendor unless you carry on these negotiations. You can negotiate only when you occasionally do business with outside vendors.

Suppose the profit centers are allowed to work with whomever they wish. The loudspeaker plant can then really buy players and receivers from outside vendors and sell them. To do that, the plant will need more resources. In such a scenario, Sound-Soul has actually been divided into four separate and independent companies. Is it more profitable to own four independent companies that produce and sell complementary equipment than to have the centers under one company? I believe that the ability to capitalize on the synergy of the combined products would lead to much better business provided, of course, that the profit centers do not ruin those opportunities. This belief is also the core of the second principle of TOC as explained in Chapter 1.

In this particular case, accepting the claim that there is no large market for cheap disc players and receivers, and also noting the doubt of whether there is a large market for stand-alone loudspeakers, it seems that cooperation among the four profit centers is crucial for the business. The need to cooperate in order to develop new integrated sets is a good example of that.

So we are back to the conflict, and it is real. How can we motivate the local managers to initiate moves for the sake of the global organization? How can we motivate them without assigning profits to whatever they do. Or, how can we assign profits and still enhance the cooperation among them.

Let us once again check the hidden assumptions behind the CD' arrow in Figure 7.2. The main assumption is that the profit centers cause the managers to look only at the profit in their own area. That is because of the transfer price system, which dictates that whenever a product is delivered to the next stage, the profit center gets the transfer price, no more, no less. When one gets a fixed price and there is no real need to ship on time and the true measure is profit, it opens the way to giving the internal products a low priority compared to the more profitable external opportunities.

That is exactly what happened with the new loudspeaker line. It is probably not the best product in view of Sound-Soul (well, we have only the accountant's word for that, but it could be true), but it is the best product for the loudspeaker plant. (The testing constraint may shift it, but we do not know the numbers and Roman, the plant manager, had not heard about throughput per constraint unit.) Giving the new line higher priority seems right from the plant's point-of-view but not from Sound-Soul's.

A New Idea for Having Both Authority and Cooperation

Suppose another system of internal payment is structured based on the idea of partnership in which every profit center gets a share of the generated throughput. The share may be different according to the product/service sold. Even if one of the profit centers generates throughput by itself, it will need to give part of it to the other profit centers.

That structure eliminates the conflict (evaporates the cloud) because it allows for a certain kind of profit center to exist and thus motivates the managers to do more and at the same time pushes the idea of partnership and cooperation. The appropriate treatment for such an idea is to create a future-reality tree. This tool has three stages of use: First, show the benefits of the proposed idea. Second, show the damage that the proposed idea might cause (negative branches). Third, search for insights (injections) to overcome these damages while keeping the benefits.

The first stage of a future-reality tree is given in Figure 7.4.

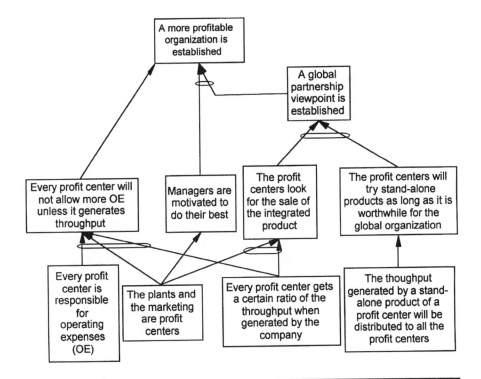

Figure 7.4   The Future-Reality Tree of the Idea About the Partnership Between Profit Centers (Only the Benefits Are Shown)

An obvious negative branch: Fixing the profit center's share in the generated throughput is not easier than fixing a transfer price. Although that is quite true, it is claimed that if the main measurement is not the annual profit of the center but rather the ratio of the generated profit to that of the previous years and comparing that to the same ratio of the whole company (meaning how much profit the company made this year relative to previous years), then this is a fair measurement.

Another insight could be to fix only very few different throughput ratios (the share the individual center gets from the generated throughput.) according to the apparent contribution of the center to that family of end products. If there is no huge difference between product families regarding the relative part of each of the centers in the throughput generation, then one ratio/share is enough. Also, when the throughput ratios are determined it should be checked that the total throughput attributed to every center, based on last year's performance, should cover all the local operating expenses and leave the same margin relative to the operating expenses. In that way, the current state is maintained, and from now on it is up to the centers to improve, either by using excess capacity to generate more throughput or by reducing cost. Please note: By generating throughput, all the centers improve. But they need to coordinate their forces in order to materialize the idea.

Reducing cost is done per profit center. A serious negative branch is that a profit center manager might concentrate more on reducing cost than on generating more throughput. The risk here is that by reducing cost, a center might be better off than by expanding throughput for which the center gets only a fixed share.

The main idea for preventing managers from cutting costs too much is to create pressure by the other partners. Suppose high-level nominations have to be approved by all the partners. This structure should lead the profit center's management to be careful not to disappoint the other profit centers. Isn't that what Aaron is looking for?

When we come back to Aaron's question: Was the reorganization into profit centers a mistake? The derived answer is yes. The way it was implemented, the organization as a whole lost more than it earned. However, instead of going back to the previous state, it makes more sense to fix what is causing the policy constraint: the measurement system. The trouble does not lie in giving the local managers much more authority to initiate their own moves. It lies in the way those moves are going to be appreciated. Change that, and the current constraint is eliminated. What will be the next constraint? It might be the testing at the loudspeaker plant. It might be somewhere else, because once cooperation is established, the whole product mix may change. And it might be that the market opportunities may be such to immediately elevate the testing constraint and move forward.

Dealing with Aaron's question, we came to realize what blocks the organization from doing more. Once that is solved, the organization will be blocked by something else but at a higher level of profits.

The above idea about partnership to evaporate the core conflict is just an idea. The purpose of bringing it up is not to claim that this is a universal solution to the profit center's problems. The purpose is to show that there are other solutions. Every new proposition that has some face value to it should be carefully analyzed. The future-reality tree is the best tool I know for carrying out the analysis. The insight needs to come from us.

This story went through a major popular trend in order to show it is wrong. It also went on to try and make it right. Whether you see the potential in the proposed idea or not, the main message is to carefully check those popular ideas.

Generally speaking, they all stem from a real need. Sometimes they cause a lot of trouble. If you are disappointed with the results of a move that was so painful to implement, use the opportunity to learn the cause and effect that you failed to see a priori. Unfortunately, not many organizations do that. Instead of checking why TQM eventually disappointed many organizations, some are throwing the baby out with the bath water and going back to old habits that were bad to start with. A pity. There are some great things in TQM as long as you are aware of its shortcomings. If you got the idea about realizing one's mistakes and then thinking afresh of what to do now, then the story achieved its objective.

8 The Profit that Came from the Wrong Product

he next story is different altogether. Instead of a problematic situation in which the reader is expected to pinpoint the core problem, all you get here is a riddle. I believe we have tremendous opportunity to learn from surprises. Yes, surprises rather than mistakes are, in my mind, the true trigger to learning. Of course, behind any surprise there is a mistaken concept or a paradigm that should be updated. So, here is a story about a pleasant surprise and the search for the wrong concept that has created the wrong expectations.

The Profit that Came from the Wrong Product—Case

I met Harry Schmidt at an executives club at the airport. His mind was clearly somewhere else. "Something wrong?" I asked him. It took him a few seconds to get away from his thoughts. "No," he said. "Everything is quite all right. As a matter of fact, it is almost too good to be true, which is puzzling to me. I simply cannot reconcile with a success that I don't quite understand. Is it normal?"

"It is common sense to me," I replied. "It is threatening not to be able to explain significant effects. Even a surprising success may repeat itself in a much less agreeable way if we cannot find a valid cause for it."

"Are you a professor or a consultant?" he laughed. "You cannot be an executive. You wouldn't stand the tension of having to act in a world that is unpredictable, irrational, and where you sometimes cannot come up with any explanation even *after* the fact.

"Well, I'm the president of a consulting company. I know the feeling, and you certainly aroused my curiosity."

He looked at his watch. "I have one hour to spare. If you are curious, I'm going to tell you the story. But don't expect any business from me. I do not use consultants on principle."

He needed two minutes to sort the story out so that the question would come out loud and clear.

"I own a family business of more than $20 million annual revenue. We manufacture mattresses. Our main line is mattresses for hotels, but we also do mattresses for homes. The year before last, we had less than $100,000 profit before taxes. I'm not complaining. It is *much* better than losing money. I have more than 220 employees whom I'd really hate to see looking for a new job. On the other hand, I know of more successful businesses than the one I'm managing.

"September 10 the year before last, I was interviewing a young kid, John McKinley, who had just finished his master's in engineering. It was a courtesy interview—I used to know the kid's father. But we didn't need a new engineer. However, the interview turned out to be different from what I'd expected. The kid started to ask questions that were directed at me: Why aren't we selling outside the state? Are the transport costs really so high, or are we afraid of the competition? Why aren't we diversifying our products? How come a reputable company like ours has only 25 percent of the market segment of the hotels in our state? What is the appeal of the other manufacturers? What other product families can we make? Are we really utilizing the core competency that exists in the company?

"On any other day, I'd simply tell the kid to go away. However, that particular morning I had wondered about some of those questions myself. What blocks Schmidt & Sons Inc. from making more money? I was quite sure it wasn't quality or due-date performance. Our competitors aren't better than us. However, there are all kinds of fashion and buzz words in the mattress business. What is the best mattress to sleep on? All kinds of techniques claim to have the ultimate answer. I know all of them. Most of them are real bullshit. But some hotel managers think it'll bring them more customers. Some of the big chains dictate to the hotels to use only the chosen mattresses. So we do have real problems. But we're a small manufacturing organization. Couldn't we be more flexible than the big ones, more responsive to our clients wishes, whether they are right or wrong?

"With these types of thoughts, the simple but bold questions of the kid had caused me to say, 'OK, wiseguy. I'm going to give you the greatest challenge of your life. You have a job for six months, at the end of which you come up with a new product line that can be safely produced with our current manpower and machinery and has a certain market we can easily attract. And don't forget the most important point: it does have to be *profitable*! That means, kid, that with that new product of yours, we'll make more money than without it. If you make it kid, you have a grand future at Schmidt & Sons or anywhere else you choose to work.'

"The kid was genuinely startled. He turned pale, but then pulled himself together and said, 'We have a deal, Mr. Schmidt!'

"As you are a total stranger to me, I can tell you something about presidents of companies, no matter what size. It is not easy to retreat from something you have just said aloud. It is simply not done. You have to look as if you know *everything*, and in no way can you be *wrong*, God forbid! I looked at young John McKinley and thought I must be out of my mind. We didn't make a small profit just to spend it on a young engineer who thinks he's brilliant. John wasn't even an MBA. I know what industrial engineers learn, and this is not what he was required to do.

"John tried his best. At the end of the six months, he appeared at the executive meeting with a nice presentation. Only he had chosen too small of a product and not at all profitable. Believe it or not, he came up with the idea of producing armchairs! He thought of buying the wood construction from a nearby carpenter and having us make the upholstery. As an industrial engineer, he had designed the production process quite correctly. It was certainly within our capabilities. He had checked the capacity profiles in all the relevant work centers and claimed there was enough capacity to generate $1.5 million in revenue. He thought this was a breakthrough. Out of that $1.5 million, more than half is the cost of the wood construction. Then you have to purchase all the other materials. And it consumes quite a lot of hand work. We would have to pay a substantial amount of overtime. Well, these kinds of workers maneuver their way to overtime no matter how much work they have. I cannot really blame them. They need the money.

"Anyway, Stanley only waited for this chance to attack John's plan. Stanley is our comptroller, and in no time he spotted that the kid hadn't calculated the true costs correctly.

"After carefully allocating the full direct cost for that product line, only a very small sum remained to cover the indirect cost allocation. Stanley's analysis came to a loss of 5 percent on the armchair sales.

"Nancy, our marketing manager, came to John's rescue. She said the hotels are interested in placing armchairs in the larger rooms, and the ability of Schmidt & Sons to supply armchairs could be the deciding factor to close the deal. Stanley just said, 'How many deals do we need to cover the loss from the armchairs' production and cheap market price?'

"John tried to ask how come others make money on armchairs. I'm the only one who heard the question. I heard it because that question bothered me. Stanley must have been exaggerating in his analysis. Anyway, I was in an embarrassing situation. All knew I brought the kid. To admit that this was a mistake was something I didn't want to happen. I kept a poker face until the end of the discussion. Then I decided to back up John's plan. This was quite a surprise to all. I enjoyed the startled reactions. They thought I was probably nuts. Maybe I was.

"In retrospect, I think I needed some change to be introduced into the company. For too many years, Schmidt & Sons had produced exactly the same products, using the same procedures and policies. It wouldn't have bothered me had we been making a lot of money, but this was not the case.

"So, the new line of armchairs was introduced into our activities. Nancy has reported that the customers have generally been positive about it. Nobody was really enthusiastic about it, but the target of $1.5 million sales of armchairs was easily achieved. Actually, it totaled more than that.

"Throughout the year I was dead worried. We have lost some customers and gained some new customers. No breakthrough was noted. Our sales looked very similar to last year except for the new product line. Based on last year's break-even state, I feared this year there would be a substantial loss due to the wrong product I had helped to launch in our product mix. Everybody would have guessed the real cause of the loss, and I could only have blamed myself. That kid only did what he could, and you cannot expect a young graduate to hit the right thing without the help of life experience.

"After the second quarter, Stanley speculated that we should be about break-even again. But I knew the last quarter was the most crucial, and more armchairs were supposed to have been sold then. I didn't rely on Stanley's analysis. I feared he was wrong.

"And wrong he was. I was wrong, too. It seems we have made a net profit of close to $700,000. You may claim this is not the world record, but this is much better than what we had done in the last five years. And it's seven times more than we made last year!

"Can you understand my unsettled state of mind? Of course, I'm pretending not to be surprised. To all my people, I say I knew it all along. Truth is,

I don't have the faintest idea what happened. By and large, we sold the same quantities of mattresses as the year before last. Bill, my production manager, takes the glory on himself, claiming his scheme to improve the setup times caused the change. Stanley backs it up because the efficiencies were up this year. I don't buy it. If we sold the same quantities, got the same prices, have exactly the same workforce, and paid the same amount of overtime, can you tell me how come we're making more money? The only additional sales were those lousy armchairs. Are the armchairs the *right* product after all? You tell me. Can it be that a product that takes more effort to build and brings in less money, is a *profitable* product?"

Just then, the announcement for the Harry's flight was made, but the old man stayed where he was. He wanted somebody to solve his riddle. I could certainly do that. What he had told me was more than enough, but I have my own business to run.

"Yes, Mr. Schmidt," I told him. "We do have the expertise to come up with a detailed explanation that would help you design your most profitable product mix. Would you like me and two of my best people to come to your facilities? It will take us two days for the three of us to give you the kind of document you really need."

The Profit That Came From the Wrong Product—An Analysis

I like Harry Schmidt a lot. He does have a lot of common sense, and he is fully aware that there is a clash between his own common sense and some of the sacred "truths" in management knowledge that is taught in so many MBA classes. He tries to find his way in this ambiguity until the conflict is too large to bear. Even then, he does not cheat himself that he has the answer. Though some doubtful solutions are suggested by his people, he is sincere enough with himself to realize that those are not the answers.

I bet that deep within himself Harry knows the answer. But its clash with the management norms is such that he does not dare to fully realize it. For that, he needs help. Maybe this is the real added value consultants can give to really good executives who are aware of the problems, sense the solutions, and badly need some backup to do what might seem crazy but is the right thing.

Before trying to answer Harry's question—"Are the armchairs the *right* product?"—let me touch on the subject of management authority. Harry points out a personal dilemma of a manager who is not absolutely certain what to do.

In Figure 8.1, the assumptions behind the arrows are part of what is called the organization's culture. These are the mental models upon which the manager is judged by his subordinates. Should a manager have very high self-esteem or rather be someone who is fully aware of the possibility that he might be wrong? The organizational norms have a great impact on the behavior of managers. Harry is certainly concerned about it, so the only person he can seriously consult with is a complete stranger. Even then, Harry is on his own. The consultant is unwilling to provide consultation for free. Harry will have to face his own principles and decide whether he deals with the problem by himself or calls for help.

Back to Harry's question: Is the new product line the right product? It is not possible to answer the question before clarifying what we mean by "right." If the right product means profitable, then it is. The added profit the company has had is certainly due to the new line.

If the right product means that the future strategy of that family owned company should rely on armchairs as a prominent line, then the answer cannot be determined from the available information.

To gain some understanding of Harry's company, let us run through the basic questions. The goal of the company is to make money. Schmidt & Sons

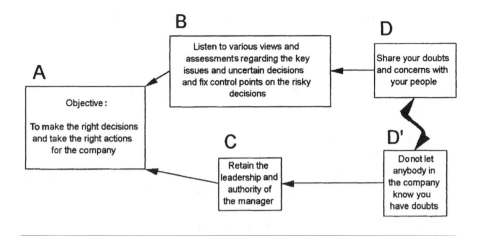

Figure 8.1   Harry's (and Others) Conflict As Managers

is a regular business organization. Can the company support all its market demand? Definitely; we know there is enough excess capacity to accommodate the new line as well. That means it has enough excess capacity all over the place. If this were not so, Harry would have told us about some effects that would indicate the possible existence of a capacity constraint. For instance, a large amount of work in process that is increasing tension between certain people and departments and lousy due date performance. The only effect that *might* have been interpreted as a sign of lack of capacity is the overtime hours. As Harry himself understands by his keen common sense, the overtime does not prove anything. It is a fairly stable number.

What we learn from the figures given is that the raw material costs are somewhat more than 50 percent of the income of the armchair line. Assume that the total throughput due to the armchairs is one-third of the sales. Because the total sales of the armchairs were higher than the predicted $1.5 million and no precise number was given, assume it is $1.8 million. The throughput generated by the armchair line is about $600,000.

Are the Direct Labor Costs Relevant At All?

No, they are certainly *not* relevant. The armchair line needs mainly manual work. The workers have a certain number of paid hours they need to fill. The regular products do not supply them with enough work. So, the workers know how to maneuver their way to overtime. With the added work for the armchair line, the overtime comes in handy, but it does not necessarily add more hours.

What Other Costs Might Be Generated By the Decision to Go on With This Line?

The truly variable expenses have been included in the throughput (T) calculation. What was not included are those expenses that incurred because of the decision to produce armchairs and are not fully *variable*. The term "variable expenses" is sometimes misunderstood. The question is, "variable" according to what? The throughput measurement was devised to include those expenses that occur per unit of sale. It certainly does not mean that they are not variable with time, for instance. When we speak about a decision that generates many more sales, we should check what additional expenses have been incurred. Those kinds of expenses are not usually incurred because of just one unit of sale. But when large quantities are involved, those expenses are not held fixed.

What Expenses Can We Think of That May Have Been Incurred in This Case?

Certainly the salary of John McKinley, the kid, should be attributed to that. Other expenses may include a rise in transport expenses.

Without the numbers, we can only guess. I assume John's salary as a new engineer would be around $40,000, and the rise in transport due to the additional line would be $20,000. Based on those assumptions, how much did the armchair line add to the bottom line of the mattress company? Remember that the profit of the company is the total throughput (T) minus the total operating expenses (OE). The profit contribution of that armchair line should be: $\Delta P = \Delta T - \Delta OE$, where the small "$\Delta$" denotes the portion that is due to the line.

We have concluded that the additional net throughput is about $600,000. The extra expenses (ΔOE) that should be deducted come up to $60,000. All in all, that armchair line has added $540,000 directly to the bottom line. The rest of the profit must have been achieved by the regular operations. Because of the small ratio between the profit and the turnover, a difference of between $100,000 and $160,000 can be caused by very small changes in the product mix.

One might get the wrong impression that the armchair line generated more profit than the whole mattress product mix. That is not true. The added profit of $540,000 was generated by the decision to add that line to the existing products. It does not mean that it would have made this added profit on its own, without the other products that consume the current capacity. It does not even mean that if some lack of capacity in a certain department is observed, we should have reduced the sales of another product to keep on with the new line of armchairs.

The additional profit has been generated by *all* the product mix utilizing the current capacity. We still do not know whether the armchair line is "better" than any other product line. We also do not know whether expanding the sales of any of the current products would have yielded even higher profits.

What should Harry do to better design the future of his company? The kid's questions are a good guide. Sales can be expanded to other products and/or more market segments. The risk in doing so is the possible emergence of a capacity constraint. Of course, once that happens there are two possible lines of action. The first is to elevate the "almost constraint" on the spot. That is not always possible and not always good common sense when the investment is high relative to the additional market. But many times that is what the company should do. It happens when a constraint emerges that should not be a constraint in the first place, either because it is cheap and simple to add capacity or because it is very difficult to control.

The other line of action is to stabilize the internal processes around that constraint—exploit and subordinate, at least for a while.

In any case, Harry should install a control system that is capable of identifying the emergence of a new constraint. The story does not tell what might have happened if a capacity constraint had emerged. That would have been another ending to this nice, optimistic story. Buffer management is a control technique capable of tracing the emergence of a new constraint. To learn more about this control mechanism I suggest *Re-engineering the Manufacturing System, Applying the Theory of Constraints* by Robert Stein. Even without implementing buffer management, being aware of the possibility of a capacity constraint may be enough to spot it on time. Simply speculate what might be the actual effects on the floor when that happens: the accumulation of inventories in front of a specific work center, the higher load spotted by the MRP (the standard production information system) reports for this particular work center, and the intuition of the production manager. When you know what *might* go wrong, you know the signals to look for when it really happens.

*There are two messages in this story that are of special importance. First is the fallacy of cost accounting that discourages so many companies from capitalizing on excess capacity to gain more business. The financial effect of doing it can be **enormous**! The second message is a subtle one, and I hinted to it in the introduction to this story. This is the need to learn from surprises! To me, surprises are the gold mine for practical learning experience.*

Crocodile Eyes: The Failure of a Great Project

W e continue here with another story about a part of an organization. *This time it focuses on a certain project that is perceived as a failure. Projects do fail from time to time. As in any other failure, the question is why did it fail and whether we know now more so that if it happens again, the result will be much better. In other words, what can we learn from this particular project? And, by the way, did the project really fail? According to whom? This is all part of the riddle.*

Crocodile Eyes: The Failure of a Great Project—Case

Whenever a novel project of development is approved, Ernest Boor, president of Rand, says, "And be careful that it isn't like Crocodile Eyes." Every new engineer who starts working at Rand hears the story about Crocodile Eyes. Somebody even dared to ask the president what he meant when he warned them to be careful of that. Ernest banged on his desk and said, "That it won't happen again, no matter what." There is no doubt many people at Rand still have traumatic memories from the project named Crocodile Eyes. Jim Morrison, who headed that unlucky project, does not work at the company any more. Jim is not exactly to be pitied; he is working at Scientist Pro as the manager of especially difficult projects. It seems that until today, nobody has been able to clarify what was wrong with Jim's actions.

The idea that stands at the base of the whole system called Crocodile Eyes was brought up in 1988 by Morton Ganes, who had developed the idea and spoken about it with the right people in the navy. The response was enthusiastic, but further inspections showed that technology was not yet ripe for the subject. Two years later, Morton had a conversation with Jim, who was at that time freshly appointed as a senior project manager. Jim said the idea was right for 1990. The technology was there, and there was no reason why it could not be done. Morton got excited, went back to the admiral in the navy, and got the contract.

Rand's management knew from the beginning that this was a novel project that included the merger of brand-new technologies and the development of those technologies for new needs that were not intended for those technologies. Jim talked about it a lot and claimed that Rand should be strategically interested in getting knowledge and experience in the new technology. Ernest, at that time still vice president of marketing, decided the project had to be profitable on its own. The contract with the navy guaranteed the sale of two systems. The contract also presented an option for the purchase of four additional systems. After discussing it with the commanders of navy intelligence, Ernest agreed to take all six systems into account when calculating the profitability of the project. Ernest also demanded that Jim performed a feasibility check. That was partially done toward the end of 1990. Ernest gave them his blessing to the project.

The project, called Crocodile Eyes, included a platform with unmanned navigation and steering abilities. The platform consisted of several instruments used for the reception and transmission of information. The project was divided into five separate subprojects, and the integration among them was done by computer software (one of the five subprojects). There was also a substantial component of software in each of the other four subprojects. Every subproject was a complete system, and the development of each consisted of many different tasks with various needs of expertise and specific purchased components. The largest subproject was the platform, which also comprised the real novel component. A special team was assembled for this specific subproject. The team included 17 people, 12 of them academics.

The development period, until the final approval of the prototype and the transition to production, was supposed to last two and a half years. The expected final date was July 1993. Every subproject was carefully planned, all the tasks specified, and its expected duration was based on the evaluation of high-level professionals, whom Jim had gathered for the project. Jim supervised and finally approved the planning of the PERT and the Gant graphs.

The development of the platform, subproject number one, was supposed to last for two years and two months, from January 1991 until the end of March 1993. Since most of the budget was earmarked for 1991 and 1992 in equal portions, the decision was to start two of the other four subprojects as well. Subproject 2, considered to be the easiest of all the subprojects, was to be finished by January 1992. Subproject 3 was planned to end in July 1992. The other two subprojects had been planned to start in January 1992, and both should have been finished by the end of March 1993. At the beginning of April 1993, after the completion of all subprojects, the plan was to start with the integration phase, which included elaborate testing of the prototype as a whole system.

Because Jim viewed the development of the platform as the main and more complicated core of the project, he asked Morton to closely manage the other four subprojects while he supervised it all, with a special emphasis on the platform. The development of the platform encountered multiple obstacles. In the field, it turned out that the feasibility check had been superficial and had not identified the main difficulty hidden in the intended size of the platform. Both Jim and Morton claimed later that it was practically impossible to perform a feasibility check aimed at the difficulties that stemmed from the size of the platform.

The first significant milestone in building the prototype of the platform was supposed to determine the resistance of the platform to pressures at high velocities. The testing should have been held in November 1991. It was actually held in February 1992. The testing brought to the surface a list of problems that shed doubts on the initial specifications. Jim told Joshua Bell, vice president for special projects, about the difficulties and asked for clearance to continue with the project. At that time, Ernest Boor had already been appointed president of the company. Joshua, seeing that the new president was the one who had given the project the initial clearance, gave the green light. Jim also asked for two more engineers and another programmer for the platform system. He was given only the two engineers, so he had to borrow one of the programmers working on subproject 3. That led to a quarrel between Jim and Morton. That small dispute was settled after Jim agreed to return the programmer to Morton at the end of September 1992.

After the whole project was reviewed again, the planned final date was to be February 1994. There was also a need to increase the budget because of the findings of the testing. Those findings added time and more purchased components to the initial planning. Morton, following Jim's orders, got assurance from the navy officer in charge that the changes in the specifications would not damage the performance of the system.

Two of the other four subprojects were developed parallel to the development of the platform. Subproject 2 ended at the end of 1991 as planned. Subproject 3, was problematic from the beginning. It became clear that some of the elements that were attached were not suitable. And it took valuable time to purchase the suitable elements for the task. Actually, subproject 3 ended only in August 1993! Morton was called once again to Joshua, who had decided to personally supervise the project in order to account for the frequent delays.

The development of the two additional subprojects started in January 1992. Again, Morton was not pleased. The development dragged on, and it looked as if it would end in the middle of 1994. The delays alone necessitated an additional budget of 40 percent! The navy refused to increase the sums in the contract, but still, considering the expected revenue from the six complete systems, the project was supposed to return the total investment and even slightly more.

The biggest blow to the whole project happened December 23, 1993, when the experimental platform was destroyed during the experiment, ruining much of the equipment of the prototype system. What quickly became evident was that there were serious doubts about whether the new technology needed for the platform was ready for the task. The lost components added to the shortage of funds, and they needed to be explained to both Joshua and Ernest. There was no way to deny the doubts regarding the technology as a whole. Even Jim, the most enthusiast pioneer of that particular technology, had to admit the difficulties in using the technology.

At that stage, Joshua, with the backing of Ernest, decided to terminate the project. When the long-faced Jim announced the decision to the development team, he justified it by saying that he did not have any solution for how to control the new technology to achieve the needed performances. One of the young engineers stood up in front of the whole team and claimed that the platform could be completed with a certain replacement technology. That announcement caused quite a provocation. Jim and Morton first dismissed the idea, and the meeting was adjourned with the decision to stop the project. But later they had second thoughts. It took Jim and Morton two weeks to thoroughly check the engineer's suggestion and find it completely feasible. Moreover, there was a good chance of completing the platform by the end of 1994 without adding to the budget! An emergency meeting with all the people connected with the project was held January 28, 1994, and they decided to breathe life back into Crocodile Eyes. There was a need to reduce the development team because three engineers had already been assigned to different projects. In general, the pressure on the performing team was unbearable because the people working there were the "crème de la crème" of the company and were wanted in other places. But the morale of the teams was high because of the challenge "to do the impossible," complete Crocodile Eyes.

The development of the platform with the alternative technology advanced as planned. But because the capabilities of the new technology differed from those of the original one, there was a need to change the dimensions of the platform. The integration tests started in December 1994. The first thing the testers found out was that due to the alterations in the size of the platform, subproject 2 was not suitable (did not fit) anymore, and they had to plan and produce that subsystem again! The tests ended in April 1995, and still there were some difficulties in the integration, which were solved by July 1995. At that point, the integration tests went well, and the prototype of Crocodile Eyes was finally approved.

The unit at the navy was happy that the project had ended and started to activate the system. The commander in charge decided, however, that two systems were enough, and that was what the contract had said. The reason for this was that because the needs of that unit had been urgent, an improvised system had been developed by the navy in 1993, a system that in spite of the fact that it needed human operators, (Crocodile Eyes is unmanned) and had an inferior specification, answered to the main needs. It was a 70 percent system, as the commander of the unit called it. Moreover, the same commander also claimed that the combative importance of the system had been exaggerated by his predecessor and had been evaluated much higher than it really was. In his opinion, that budget could have been used for more important tasks for which there had not been enough finances.

Jim was very angry with the unit commander. He said that in the regular meetings with the unit personnel, nothing had been disclosed about the alternative system they had built.

Ernest was furious because of the failure of the project. He summoned Jim, Morton, and Joshua to a long meeting, during which very harsh words were spoken. Jim, who had understood that his future at Rand was not secure anymore and had started dealing with Scientist Pro, retorted that Ernest had turned a blind eye and had not wanted to see the difficulties. Jim claimed that such a novel project should have gone on for at least eight years. He even brought up data about an air force project, somewhat similar in the task and the technological difficulties to Crocodile Eyes, that took eight years to complete. Jim said, "There is no such a thing as a project without delays and even technological failures. What matters is that every problem finds its solution. This system is a magnificent system for any navy. We know that in the Ministry of Defense they are showing great interest in our platform for different needs."

Morton added to Jim's speech by presenting a long list of other products that could be developed very easily thanks to the platform. Morton said although he could not dispute that certain things had gone wrong with the project, it still might become profitable from using the knowledge and the systems developed elsewhere. Jim was quick to continue along that line, claiming that the rest of the equipment that had been developed for the project was better than anything that existed in the world. Jim concluded, "What matters is that in the end, the system achieved everything we proclaimed it would."

Ernest's reaction was cynical, and he said Crocodile Eyes was a colossal failure of a great project.

We must point out that the platform developed for Crocodile Eyes was later included in several projects that Rand received in the following years. It is quite possible that the bottom line is that Crocodile Eyes helped Rand increase its profit. Jim, in his position in the competing company, nurtured the same novel technology that failed in Crocodile Eyes in several novel and very successful projects. All the same, the troubled project of Rand became the reference for a poorly managed project. It is certainly something for people to learn from.

Crocodile Eyes: The Failure of a Great Project—An Analysis

We all make mistakes all the time. What makes this story a sad one for me is the obvious reluctance of the principal characters to learn from their mistakes. It is not to be expected that Ernest Boor and the other managers at Rand learn all the insights Dr. Goldratt presented in *Critical Chain* from their own experience, but some learning should have occurred. As it appears in the story, no learning or even rethinking about the decisions made took place. In my opinion, rarely does a failure like the one described initiate true organizational learning. I assume *some* personal learning took place. I can only hope that Jim Morrison, obviously a fairly good project manager, learned something and that he will be more cautious in similar circumstances. Rand, as an organization, did not learn anything. If there was any kind of learning, it was the wrong lesson. For instance, in this particular case it seems that laying off Jim because of the consequences of the project was a mistake.

I believe that learning from the experience of a particular event is initiated by a significant gap between the expectations and some actual outcomes. When there is no surprise, it is difficult to raise the question "Why?" When such a gap is recognized, it means that deep in our thinking there is a paradigm that needs to be updated. That presents a great opportunity to learn from our experience. Identifying a gap between an organization's formal expectations and actual outcomes is an even greater opportunity to learn something that will help improve the future performance of the organization.

What could have been learned in this case? We need first to identify the most painful and significant gap. In this story, several expectations were not met. Where do we start?

Let us bring up some options. One obvious gap is the time it took to build the prototype of Crocodile Eyes. It was predicted that it would take two and a half years, and it was actually completed in four and half years. Another gap is the commercial failure: Instead of selling six full systems, only two were sold. That gap can be viewed as even more serious because the expenses were considerably higher than the initial budget. A third gap is the technological failure that almost killed the project. That was not expected by anyone.

Of those three, let us start with the one most painful to Rand, which is also the most significant—the commercial failure. Why do I say that gap is the most significant? Because if it had ended with six systems sold to the navy, I doubt very much that the connotation of "colossal failure" would have

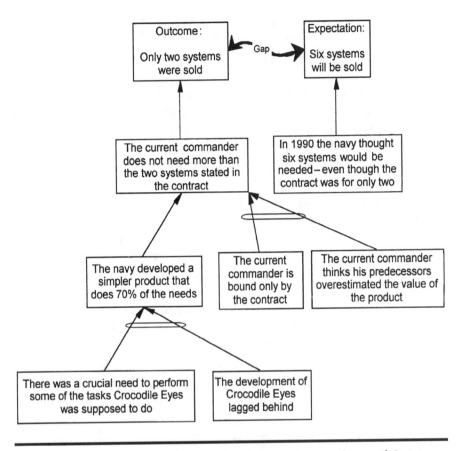

Figure 9.1   The Causes for the Main Gap Between Expectations and Outcomes

applied to it by Ernest. This observation also emphasizes the gap in the number of systems sold to the customer rather than the additional expenses. We will still see that the other two gaps will be included in the analysis, otherwise we would not be confident that we had gained the proper understanding of the cause and effect of that event.

Figure 9.1 is a brief logical tree that tries to explain the particular gap.

The gap is represented by the two conflicting effects. That is done to provide the possibility to understand how the expectations were created.

The simple logical tree outlines two explanations for the disappointing sales. One is created by the lateness of the development of Crocodile Eyes. Had the prototype been ready in mid-1993, there would have been a good chance that no alternative simpler solution would have been developed. That would have raised the probability that the navy would have asked for more than two systems. This reasoning of what could have happened provided Crocodile Eyes had been finished on time is depicted in Figure 9.2. This is a "what-if" kind of logical tree that tries to speculate the outcome of a hypothetical scenario. The other explanation is directed at the new commander's attitude.

The logical map easily reveals that because of the possible appointment of a new commander and the possibility that he underestimates the value of Crocodile Eyes, it is not certain how many systems would have been sold. So the conclusion has to be that the timely completion of the development of the prototype does not ensure the sale of six systems!

That means that something was wrong with the confidence of selling six systems. With the possibility of a change of command and the underestimation of a new commander, the value of the system was something that was not considered.

But, we can certainly say that the *probability* of selling all six promised systems was heavily affected by the lateness of the development. Here we have reached the gap regarding the time it took to develop the prototype.

Now, we come to inquiring about the causes for the delay in the project. We can start with two broad and generic hypotheses that may lead us to inquire deeper into the known facts and what we can deduce from them. Each hypothesis can explain the gap, but the existence of the effect in reality needs validation.

The first hypothesis looks at the part of the expectations of the gap. If those expectations were not realistic to start with, then that explains the surprise and disappointment from the actual outcome.

The second hypothesis simply claims that the project management was poor. That is clearly a very broad statement that needs to be further detailed.

In Figure 9.3, the logical arrows here are less precise than what is usually required in the TOC thinking processes tools. For example, the claim "the development of the platform took much more time than expected" and caused the project to end in July 1995 should be grouped with another effect that no other subproject was delayed by that much. We know that is true, and it does not seem fully necessary to put it there. The other effect, "Subproject 2 didn't fit the new size of the platform," acts independently in delaying the completion even more. Hence, the two arrows going into the actual outcome box are not grouped together.

A fact that includes numbers is difficult to explain without going into too many details. Why did they expect the project to complete in July 1993? The first hypothesis is that the evaluations were too optimistic. That statement

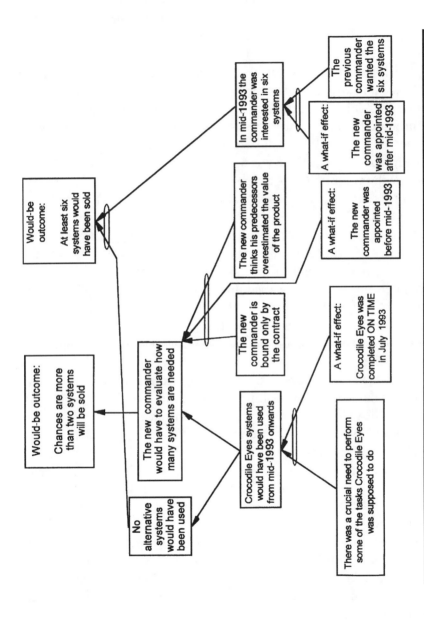

Figure 9.2   What Could Have Happened If Only the System Had Been On Time

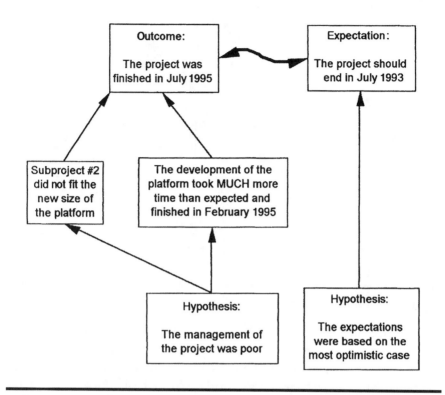

Figure 9.3   Two Broad Hypotheses Each Explaining the Gap

does not relate to the precise number, but to the difference between the expected and the actual dates. Such a logical connection, is in my opinion, good enough. It looks redundant to add an effect stating that considering all the tasks, evaluating each task's duration, and looking at the resulting total duration, the project came to an end in July 1993.

The gap is represented by its two conflicting effects. Any explanation of one side explains the gap. If the expectations were based on too optimistic an evaluation, that also explains the late completion date.

According to my own experience, we first need to put some generic claims on paper and proceed from there. Certainly the hypotheses are just intermediate steps. We need to detail and substantiate them and then even proceed to the possible causes of those effects.

Let us start with the hypothesis regarding the expectations (Figure 9.4):

The hypothesis that the expectations were too optimistic could explain the gap. It also could explain some of the other effects known. Some of the links do not specify additional necessary conditions that some purists may

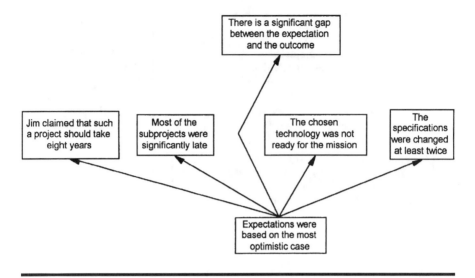

Figure 9.4 Substantiating the Causality for a Hypothesis

wish to include. For instance, the chosen technology could have been ready—if they had been luckier. In a more rigorous logical drawing, an entity specifying that the knowledge about the new technology was not developed enough should be written and grouped with the basic cause to create the effect that the chosen technology was not ready for the mission. Jim's claims that such a project should take eight years could be explained as just an excuse. But the hypothesis says it may be true, at least somewhat true. What can be seen is that it is *reasonable* that the expectations were based on an optimistic scenario because it could cause several effects that occurred in the environment.

We do not always get such support of a hypothesis. Collecting the main effects that were assumed to be caused by the second hypothesis does not seem to fully substantiate it. Figure 9.5 displays this logical map. The hypothesis of "poor management" was replaced by "Jim and Morton are not good project managers." The reason for the change is that any cause for the gap could be described as "poor management." The hypothesis that should be checked is whether Jim and Morton are lousy managers. That also means that there are other people who could manage much better.

From Figure 9.5, we see that if Jim and Morton had not been good managers, most of the subprojects would have been late. It also explains the failure to identify the necessary change in subproject 2 and the choice of the technology. Two of the three effects could have been caused by the first

10 Missing Information*

A *warning! This is a fast moving story for which you are required to identify the message as well as whose problem are we looking for. It is my own variation on a comedy of errors. But there are some real problems here, and people do pay the price for their mistakes. So, the question is whether they could have pinpointed the mistakes long before. That is a hint, and you may also speculate about the title of the story as a hint as well.*

Missing Information—Case

David Serry was sitting in Daniel Cane's office in Tel Aviv when Josh Shamir, the famous CEO of Fountain, announced his resignation during a press conference. Josh did not supply any further information. David knew about the press conference but thought he needed a rest after quite a hectic week. So, he was chatting comfortably with the ever-worried Daniel when the big news happened.

The conference was called to report on the financial results for the last quarter of 1995 for Fountain, the largest soft drink company in Israel. As usual, the financial data had leaked out beforehand. And it was not too interesting to David. Everybody knew that since Josh had taken the CEO position in January 1991, Fountain had started making a profit. The last quarter of 1995 was supposed to be a little less profitable than the parallel quarter in 1994, a decrease that was attributed to investments in the Soda-Wine line. In spite of all this, Fountain still makes money, a fact that makes Daniel an utterly unhappy person.

---

* Taken from *Status, The Magazine for Management,* Narkis Weinberg, Ed., 1997. With permission.

Daniel is the CEO of Shultz breweries. Ever since Light Beer had been introduced by Shultz, Daniel feared that Fountain would introduce the Soda-Wine as a competitive product. His conversation with David, the food and drinks industry senior reporter for the business paper *Cash Flow*, was precisely about this subject. Daniel and David had been exchanging information for years. David wanted to know what had been happening in the Swedish mother company concerning the rumors about opening a large brewery in Cubno, the former capital of Lithuania, in association with the German company Waiser.

Daniel was willing to deliver, but he wanted to verify the rumors he had heard from Business Investigations, an information business provider company that claims to incorporate the latest technology used in military intelligence. Those rumors had said that Josh had strongly opposed the introduction of the Soda-Wine and that the line had been purchased by Sharon Rozen-Bain, the vice president in charge of business development, during the three months Josh was in France trying to help the French subsidiary stand on its own two feet. David approved the information all right. He even quoted Josh saying to Sharon during an executive meeting at the beginning of November 1995: "Such competition is an unbusiness-like gamble. Wait for your turn to make big mistakes. I have already made mine, and I am not going to make them again."

At that moment, the way things usually happen in the large theater of life, David's cellular phone rang. It was Sammy Soran, a fellow reporter who covered the food industry with David. He told an astonished David about the new and unexpected subject of the press conference. David turned pale. This is not a position in which a senior reporter would like to find himself.

How was it possible that there had not even been a hint from Fountain about Josh's intentions? Something obviously had gone wrong. William Gold had written two months previously in the *Financial Times* that Josh and the board of directors had differences of opinions, but Gold was not highly thought of in the field and his sources were not very reliable. You also did not argue with such clear success. Josh and the board of directors had already had differences at the beginning, back in 1991, when Josh decided to stop the pineapple juice line and center on the citrus and the strawberry-banana juices. He cut back the variety of items from 40 to 28. The board gave in and was very satisfied. David was sure that Gold did not realize Josh's status in Fountain.

At the beginning, Daniel felt immense relief. Josh was a dangerous adversary, and it was good to see him leave. True, the competition between Fountain and Shultz is only marginal—soft drinks cannot be a real competition to beer. But it is a fact that the chairman of the board of Shultz diligently collects material about Fountain and keeps asking Daniel why Shultz is not as efficient as Fountain. This comparison seemed very unfair to Daniel who feels the beer market in Israel is much more difficult to deal with. Israelis have been picking up on drinking beer, but the average consumption is still relatively low.

Then David asked, "I wonder who will be the next CEO. If it is Sharon, things will go wild in Fountain."

Daniel's heart skipped a beat. Sharon seems like the natural choice. A fierce competitor to Josh, she was the one who promoted and fought for Soda-Wine. The gossip columnist of *Cash Flow* had already written about the "special attention" Sharon got from Edgar Blumenfeld, Fountain's chairman. If Sharon is appointed CEO, Soda-Wine is on its way. That means Shultz would have to launch the extended campaign for Easy Beer. It was seriously feared in Shultz that sales of Soda-Wine would be partially at the expense of Easy Beer. The campaign was ready and waiting for Fountain's next steps. The primary investment in the campaign had already been made, but launching the actual campaign would mean an additional expense of 6 percent in the advertisement budget. And it would only allow sales to remain at the same level and not to lose sales to Fountain.

The more Daniel thought about it, the angrier he became with Business Investigations. One does not expect such a surprise when so much money is paid to get this kind of information on the competitors. Forgetting that David was sitting with him, Daniel picked up the phone and called Dan Sallinger, director of Business Investigations.

The conversation was unpleasant for both parties. Dan repeated again and again that the information Business Investigations had been asked to supply was about "new products and marketing policy" in Fountain, as well as "any relevant financial information." He claimed that they had concentrated on these topics. The information about the experimental development of Soda-Wine had been given personally to Daniel a year before. It had been given with the additional evaluation that Josh was not keen on the idea, and this was an understatement, but had avoided an open confrontation with Sharon, who had started to gain the support and sympathy of Blumenfeld. The investment of $3 million in the production line is included, to the last detail, in the reports that were given to Daniel, including the field test that had been held in January and had been described in internal reports as "quite successful." Daniel was not impressed. The report about the Soda-Wine line was published in *Cash Flow* in November, so it was no secret. How was it that it was not reported that Josh would retire or "be retired" from Fountain? Wasn't this information part of "any significant financial information?"

The tone of the conversation rose, and at the end Daniel said he was severing any further business deals with Business Information. He also said the information about the market survey of Krops Beer was not essential, and he was ready to renounce it. The main reason was that Gross-Sweets, the company that was believed to be in negotiations with the British company to get permission to manufacture the beer in Israel, did not know enough about the drinks market in Israel. With all due respect, you have to work hard to acquire that kind of knowledge. Moreover, Michael Dover, the deputy general manager of Gross-Sweets, was a young fellow who still had a lot to learn. If Business Information could not differentiate between important information and simple data, it was not needed any longer.

David was sitting and writing. This was the compensation for the huge mistake, he thought, something that might take his editor off his back because of the shameful miss. Still, Josh's resignation had been an unpredictable move and an impossible one to foresee. In 1991, it was David who had first foreseen that Blumenfeld would bring Josh to Fountain. Blumenfeld and Josh were both, each in his own way, difficult people to deal with. Josh said Blumenfeld did not know where the bottom line was because he remained stuck in the expenses articles. When Blumenfeld was told about this remark, he asked, "If this is the case, what is he doing here? This Josh is not exactly the miser of the year. But as long as he brings in the profits, that is what really matters." What happened then? Josh continued bringing in money. Was that the time for such changes?

What mattered now was to guess correctly who was going to be the next CEO. Sharon was Blumenfeld's best gamble. But he could not afford letting Max Bishop, the chief financial officer, leave, too. Max wanted to be the next CEO, and he was not one of Josh's fans. Actually, Josh's era in Fountain was over. Two years before, they had talked about Tom Kessler as "Josh's apprentice," but Tom had left the country six months before and seemed to be doing well with Philips. He did not need Fountain anymore.

It was time for David to plan an article about the end of Josh's era in Fountain and leave an open question: "Where is fountain going, forward or backward?"

Thinking about "Josh's era" reminded David of the chance meeting he had with Josh and Michael Dover at a wedding. David had noticed Josh talking with Michael Dover, CEO of Gross-Sweets. When he had approached them, Josh said, "Hi, David. Do you know Michael? A bright mind. He is only 31, but he has a great future. And I am not saying this just because he is my cousin's son."

The press conference in Fountain was held on March 3, 1996. On March 29, 1996, Michael Dover called a press conference in Gross-Sweets. The rumors that spread during the couple of days before the press conference were about an agreement between the British brewery Krops and Gross-Sweets. The agreement talked about producing beer in Israel. At the staff meeting of *Cash Flow* on March 28, 1996, Sammy Soran reported that there were rumors about a "Turkish Connection" to this agreement. This information looked dubious on the face of it, and the staff decision was not to relate to it in the meantime.

David and Sammy both attended the Gross-Sweets press conference. During the first part, Michael presented the details of the agreement for building a brewery in Dimona, which is in the south of the country, with the government's aid, which had been agreed on several days before. The brewery would start production by the beginning of 1997. Meanwhile, one of the subsidiaries of Gross-Sweets would import Krop beer from Turkey. The modern Turkish brewery had opened only a few months before, but the marketing forecasts had been wrong as a result of the shift of power in Turkey in favor of the Muslims. The Turks loved the idea of producing for export during 1996 with the full approval of the mother company from Britain.

During the second part of the press conference, Solomon Wain, chairman of Gross-Sweets, presented the new CEO of Gross-Sweets: Josh Shamir. Michael Dover would go back to being the deputy but would actually manage the local branch of Krop.

"How could this happen to me?" David asked Sammy. His only comfort was that Daniel was probably feeling worse. True, Daniel had already heard about the agreement, but the Turkish part was totally missing from the information he had received from Business Information Company, which he had approached after leaving Business Investigations. Business Information dismissed Daniel's complaint without any problems: "Business information cannot be 100 percent accurate. We weren't looking in this direction, so we didn't find it," it said. The information office had known about the dealings between Josh and Gross-Sweets, but it had not been given to Daniel because he had asked only for information concerning the agreement being negotiated between Gross-Sweets and Krop.

The editorial staff of *Cash Flow* decided on an internal inquiry to check their sources in the food and drinks industry. In the Shultz breweries, an investigation was started to find out what had been causing the company not to be ready for immediate competition with Krop beer.

Missing Information—An Analysis

There are two unhappy people at the end of this story. Are they unfortunate or did they miss something? In an unpredictable fast-moving world, it is not surprising that some people are taken by surprise. Could these people have guessed exactly what was going to happen? Certainly not. So, could we conclude that this story is about luck and not at all about a management dilemma?

I'm going to try to dispute the last sentence. Acknowledging the uncertainty surrounding us, we can still focus on the most important issues and construct a protection mechanism to keep us away from mere "bad luck."

And there is, in my opinion, a hidden dilemma that has caused a substantial part of the unfortunate events for both people. The dilemma is shown in the following cloud in Figure 10.1.

The conflict is displayed in a very generic form. In the story, Daniel Cane, the CEO of Shultz, targets the potential competitors. David, the reporter, lives through the conflict in its most general form.

Let us verbalize some of the key assumptions behind the arrows.

AB: The most important assumption is that every incident that occurs might be meaningful. In a way, this assumption recognizes the partial dependency that exists in reality. Another important assumption is that there is no way to know prior to an incident whether it is meaningful or not. Hence, there is a need to know everything.

BD: The assumption here is that if you collect information, then you know.

AC: The main assumption here is that to make a decision, one needs to be able to evaluate the information at hand.

CD': The basic assumption here recognizes the inability to control too much information. It imposes a limit on the amount of information a human being can consider in the decision-making process.

The assumption that TOC challenges is the one behind the AB arrow. It also attacks the BD arrow, claiming that in order to "know," you have to be able to analyze the meaning of the information. That means what new cause and effect relationships are derived by that piece of information. If you are unable to do that, then collecting that piece of information does not have any added value.

The main assumption that is challenged is that every piece of data can be of value. Do we need to know "everything" in order to make the right decisions? Suppose a company needs to choose an ERP system. Further, suppose this is a real necessity. How many ERP systems exist in the market?

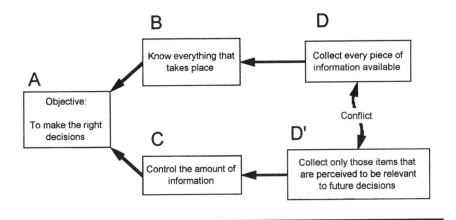

Figure 10.1   A Generic Conflict

How much do we need to know about an ERP system to decide if it suits our needs? How do we choose among those which suit our needs?

Some companies use a very lengthy process of collecting information and a sophisticated decision-making process to make that decision. Even those companies find themselves trapped in the conflict and reject some of the packages without any real consideration.

To get some control, the collection of information for an ERP system is based on a "want list." This is already a concession to the necessity to gain control on the information. These want lists tend to be very long, hence no ERP system meets *all* the requirements. Now the decision-making process is trying to set priorities and decide what carries more weight. At the end, there is some kind of "equation" that points toward the winner.

This is exactly the description of the above conflict. On one hand, there is an urge to add requirements and check more candidate systems. On the other hand, there are concessions to the human inability to measure benefits and control the amount of information.

TOC focuses on the constraints of the organization. That emphasis provides the opportunity to distinguish between something that is crucial and many others that are "chupchics" (a word Dr. Goldratt has transferred from the Hebrew slang that means something not significant). The direct search for meaningful data is something worth adopting from the TOC philosophy. The direct search starts from the goal and passes through the constraints, then analyzes the subordination processes.

Although choosing an ERP system is a planning decision, the control part of the management mission is in even more need directing the search for meaningful data. In a planning decision it is easier to focus on the needs and then develop the information needs. In a control phase the crucial questions are as follows:

1. What might go wrong?
2. How can we prevent it from happening?
3. If it actually goes wrong, what can we do about it?
4. How can we know what goes wrong early enough to take the necessary actions?

Question 4 has been put last because when something goes wrong and we can do nothing about it, then knowing in advance has no benefit. As an example I can bring up the election polls. Before the elections, the polls have a real impact on how people vote. Once the elections are over and all are waiting for the final results, all the polls and partial "information" flowing around have no impact on the decisions regarding the elections. The whole benefit is for satisfying curiosity.

Going back to the story, let us see whether Daniel and David could have focused their information needs in a different way. If neither is going to carefully study how they could know in advance what took them by surprise, they might both learn the wrong lesson from what happened.

Daniel might decide that information coming from business-information companies is no good. David might conclude to attend every press conference, to write down every family connection he knows about, and mainly to pray for better luck should be his decision.

These definitely are the wrong lessons to learn. I fear wrong lessons more than I fear not learning the right lesson.

Daniel Cane's Case Analysis

We know about two efforts of Daniel to collect information. Both are directed at possible competitors. This is part of the managerial control. The first question is what might go wrong? The simple answer is that the competitors might do something that will take away part of Shultz's market. Better verbalization of what the competitors might do could lead to a list of possible actions, such as launching a new product, a new advertising campaign, a change in the pricing and/or promotion actions.

Suppose you know your competitor intends to do one of those options. What can you do about it? Usually it is not feasible or legal to prevent the competitor from taking the action. What is left is to counter react. How do you know early enough? How early is it needed to be for the reactions? Sometimes knowing after the fact is as good as knowing a year ahead of time.

The first possible competitor is Fountain. Currently, Fountain is not a true competitor. The marginal indirect competition is not worth looking into, as nothing can be done about it. However, when Fountain considers a new product that contains some alcohol, it might present more significant competition that might call for some reactions from Shultz.

Was the search for information on Fountain well-established? Not before the news about the Soda-Wine line. Was it necessary to look for information on new products and marketing policy of Fountain? If all the potential threat Fountain posed to Schulz was a mere chance of launching a new product with alcohol, it seemed that such news could have been widely known early enough for Shultz to react. There was no need to look closely at Fountain. Why did Daniel look into Fountain? The most reasonable explanation is because of two effects acting together. First is the personal rivalry between Daniel and Josh—or rather envy, an envy that was intensified by Daniel's boss referring to Josh's success. However, the envy is not enough; purchasing information needs to have the visibility of something that is required. How come that information about Fountain is perceived as necessary? That is because the wrong principle of maximum data on anything that is vaguely relevant is quite common. This wrong principle is the second effect that together with the personal envy led to looking for information on Fountain.

Note that the purchase of the line for Soda-Wine was published in the papers only a short time after Daniel had received the report. It is very likely that such a drastic move would find its way to the headlines. Daniel did not need to look closely at Fountain. That kind of news becomes known anyway. And the preparations for launching a new product take enough time for the competitors to prepare an advertising campaign.

Suppose Daniel had received the news about the resignation of Josh earlier. Would it have posed any better decision or action? Probably not. Daniel had already taken the precaution of having a campaign ready. Was it the right action? Never mind; the main point is knowing too early is useless.

The case with Krops Beer and Gross-Sweets is much more severe. The main damage caused by watching Fountain closely was ignoring the penetration of Gross-Sweets into the beer market. That was a much more threatening move. Here, there was a basic flaw in the interpretation of the initial

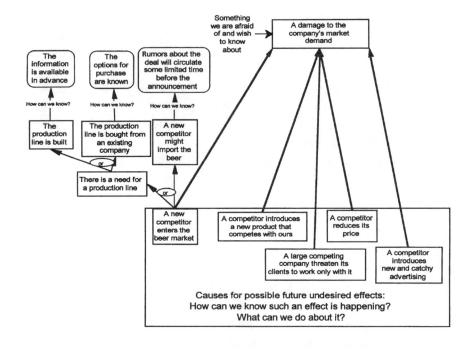

Figure 10.2   What Should Have Been Daniel's Control Tree?

information. Daniel knew about the intent. How such an intention could be translated into a fast launch of the products was something Daniel never tried.

In this case as well, Daniel was acting according to a wrong concept or paradigm—having contempt toward a young director and merely saying that the sweet company lacked the necessary experience in beer. We would expect an experienced CEO to try and speculate the moves of his most threatening competitor. All the signs were in place. The solution of first importing the beer before producing it locally is not something that should have been dismissed.

How should Daniel have planned his control system for the competition? The following control tree tries to highlight the thinking procedure (Figure 10.2):

At the top is the effect the company tries to prevent, or react against, on time. Below should be all the significant causes that might cause the effect at the top. Please note, unlike "regular" trees, here we place effects that right now do not exist (or so we hope) in reality but might be caused. What we look for is to be able to identify the emergence of the threat early enough to do something about it.

How can we know if a new competitor is going to enter our market? When such a cause occurs, it causes other effects, some of which are visible enough so we can establish the way to get the really crucial information on time.

Those kinds of control trees are required to build a good control system, where the information is collected based on known and defined needs.

David Serry's Case Analysis

David, the reporter in this story, has very different information needs. He is not looking for the big picture of the organization that managers are trying to construct in their minds. All he is looking for is the kind of news that would interest a wide variety of readers. The resignation of Josh has limited and indirect managerial interest for Daniel, but for David it is the backbone of what he does. Guessing that such an incident is possible and even expecting it could have been of great value to David. He did not guess. Did he miss something?

Oh, yes! He missed the signs. If you are in the speculation world and if guessing right is a basic characteristic of your job, then you need more than regular intuition. This kind of extra intuition can be developed and nourished by learning to ask yourself questions about the motivations of other people. Here is a well-respected CEO who has some differences with his board, but as long as he succeeds, the board members leave him in charge. How should someone like that feel when he learns that his vice president has taken the opportunity when he was away to purchase very expensive new machinery that is good only for a product that he refuses to launch? Is it enough that he says he will not go for it? Suppose the board agrees to his terms. For how long? What is inevitably going to happen when the profits drop? What can be learned from the fact that the vice president remains in the company? How long will the huge investment that has been made and that is now of no use, stay in that state?

Let us construct a logical tree of the rumors David knew about and speculate what might have happened if the rumors had been true.

The logical tree in Figure 10.3 suggests the possibility of Josh looking for another appropriate job. That is a possibility David never thought of. It could explain the effect (not appearing in the tree) that no rumor came from Fountain by making the assumption that it was Josh's initiative to leave and take the position at Gross-Sweets because he liked the opportunity and was angry about what took place at Fountain. Of course, were David aware of

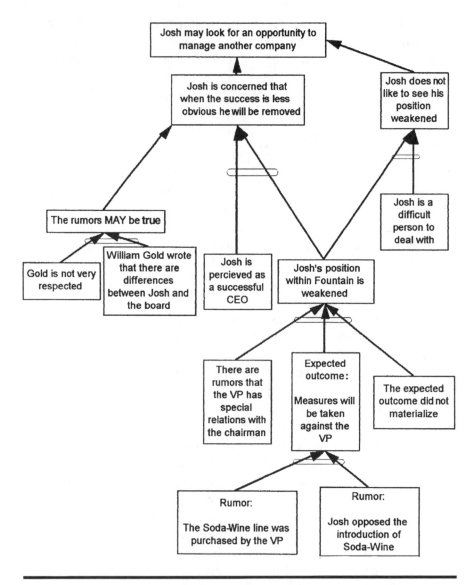

Figure 10.3   Analyzing the Data on Hand—Where Does it Lead To?

the possible outcome of that simple logical map, he would have considered Josh's moving to Gross-Sweets and might even remember the family connection between Josh and Michael Dover. This is a good example of having the data but not really "knowing."

Processing data is widely considered as computerized operations. From the TOC perspective, processing data converts the data into information.

Dr. Goldratt defines information as an answer to a question asked. Mapping the known, or even assumed, facts into predicted effects or speculating about the causes of certain known effects is, in my opinion, a major processing of data into information where the human factor, the logic, as well as intuition are necessary. So, the missing information is really missing information. The data was there; the human processing was missing. Daniel's and David's misfortune is due to themselves.

We must recognize the need to evaluate informal and formal data and derive the directed and focused information we need. That includes the control requirements where one starts with the concerns and speculates upon possible causes to identify the data that could easily be gathered. These lessons are part of the change that TOC brings with it.

The relationship between data and information is, I hope, highlighted in this story. It calls for thinking processes tools beyond what is currently suggested. Control trees are of immense importance. Every intelligent manager can expect that certain undesired effects might pop up. In certain cases, knowing as early as possible that these are going to happen can lead to the right actions to eliminate or reduce the undesired part. This recognition has a lot of ramifications. Just think of a case in which your marketing manager goes around to your clients and maybe looks for a better job for himself. Pretty scary, isn't it? What data do you look for that will give you an early warning that this is actually happening? Try to construct a control tree for that. I think this is a nice exercise.

11 Planning the Next Season and How to be Supported By the Information System

T his story is about some troubling details. The CEO of a tourist operator company is planning its next season. Many numbers that can be interpreted in several ways enter such a planning. Do the numbers tell the story? I don't think so and neither does the reputed CEO. He is in a dilemma that we should try to relieve. The notion of information, especially concerning "hard date," meaning precise numbers, is full of traps. Beware of the support of information systems unless you know exactly what the underlining basic assumptions are.

Planning the Next Season and How to be Supported By the Information System—Case

Caesar Corp, the reputed CEO of Vacations Unlimited sat at his table and cursed. It was mid-September, which meant that the annual planning had just began. It was tedious work. In a week or two, the best people and agents of Vacations Unlimited would go to the following year's vacation resorts, sign contracts with hotels, local instructors, local tourist companies, and, of course, airlines. All would be targeted at providing great, exhilarating vacation packages. Until then, Caesar needed to do the planning himself. Well, he had already had the tour packages recommended by his subordinates, as well as some analyses of the trends in American outgoing tourism and an analysis of the expected impact of the Western economies on vacation habits.

In the end, the decisions all lay with the CEO. Caesar trusted his intuition in guessing the popularity of each vacation resort each year. It got tougher every year. Of course, the usual places such as Cancun and the Virgin Islands would have the expected share. However, those places meant very moderate business—a lot of competition drove down the price. Finding some new spots seemed much more profitable. But you also needed to customize them to the level of service and length of vacation that would be sought by potential customers. One should be very careful about it. This was what the new computerized planning system was supposed to support. It was supposed to add up all the predicted expenses to provide an assessment of how much the price of a package tour should be to make it profitable. Of course, the *actual* price for any package tour was not based on adding up the expenses and showing a good profit. It was the competition that set the price.

Caesar was looking at the computer. It seemed like a nice program—good layout of the screens and a lot of features. For any suggested package tour, it filled in the costs that could be collected from the previous year. If the same package had been offered the previous year, you could easily get the previous year's forecast, actual sales, the price at which it was sold, and the actual cost of the package. Caesar could mess up the expenses by changing the assumed costs of the flights, hotel rooms, and other services. He could also change the designated aircraft and the standard of hotels to be used for every package deal.

In the end, the planning support program would be given an estimate of the costs for each package. The costs included the commissions for the travel agencies. What was definitely *not* included were Vacations Unlimited's internal costs—the computer system, the hot line, the communication, and the personnel. It was Caesar's decision to not include those expenses, as he considered them to be fixed costs. Some people in Vacations Unlimited did not agree with Caesar on that point, but he was Caesar.

The planning support information system also sorted the packages for every destination according to its profitability, allowing the decision maker to decide what to include. Several package tours went to more than one vacation resort, and the program was fully capable of handling it.

"This is a very neat software program," Caesar thought to himself. "So, how come I don't believe what I see?"

At that moment, Caesar was looking at a particular group of packages for Palma de Majorca, Spain. The program already knew the kind of aircraft that could cross the Atlantic. Vacations Unlimited rented the aircraft, the crew, and the terminal from an airline company. The cost of the flight was $51,000 for a round-trip flight, and the aircraft could carry up to 125 passengers. The average booking for this aircraft during the July to August season was 85 percent. Another problem was that at the very beginning of the season and at the very end, some of the flights were almost empty even though some combinations with regular flights reduced the negative effects. Inclusion of this factor reduced the average bookings usually calculated for regular vacation flights. Here, the mechanism offered by the program was to add a certain percentage to the cost of flights. Caesar decided on a 10 percent additional cost based on the previous year's results.

On the screen the four packages being considered were shown.

Package	Hotel Expenses	Flight Expenses	Other Expenses, Local Expense and Commissions	Total Expenses	Selling Price	Profit per Package	Forecast of Packages Sold
A	600	528	307	1,435	1,795	360	160
B	400	528	92	1,020	1,295	275	200
C	500	528	125	1,153	1,595	442	75
D	550	528	279	1,357	1,300	−57	50

The last package, D, was a group package. That meant there was a commitment to buy the 50 packages. This time it was intended for two groups of 25 people in each group. The low price was because of a counter offer they had shown Caesar. Before looking at the financial analysis, Caesar was certain that the offer was good enough. As a matter of fact, he had asked one of Vacation Unlimited's best salesmen to book another group with the same conditions. At that point, it looked like a losing offer altogether. Still, Caesar's intuition was that it was good business after all.

The forecasts that appeared on the analysis needed more elaboration. Generally speaking, they were Caesar's estimates. He predicted the following year's demand for every package based on the previous year's actual demand, the forecast of the economic factors, and some informal information about what was considered "in" and what was "out." With all that, Caesar's intuition, plus 35 years in the tourist trade business, set an expected demand. By the way, the actual demand of two years previously seemed useless to Caesar. People changed their minds fairly fast.

The first round of decisions was the choice of destinations and the type of packages.

The second round dealt with planning the flight schedule. This was done by Murray Fisher, who handled the relationships with the airlines. The first schedule was just a target—the stiff competition for available aircraft made certain dates and preferences impossible. In any case, the schedule should be fixed no later than October 1.

At the same time, the hunt for hotel rooms at the preferred resorts was carried out. This would not be signed until the third round of decisions was finalized. This was where the packages were fixed, along with the selling price and the dates, taking into account the limitations of the flight schedules and the availability of hotel rooms appropriate for the kinds of packages.

After the third round, a catalog specifying each package was printed and sent to all the travel agents with whom Vacations Unlimited worked.

Certain changes in the planning might be carried out. Some hotel rooms could be canceled up to two weeks before the proposed date. However, that should be done with caution. The number canceled could not exceed 30 percent of what had been agreed upon. Moreover, if Vacations Unlimited canceled more than 10 percent of the rooms from a certain hotel, it would be difficult for it to have a good deal with that hotel the following year. That was especially critical with the hotel chains.

Generally speaking, it was the same situation with flights. You could cancel a flight a week before the schedule, but you had to be careful not to overdo it. Canceling flights was very problematic from the perceived service point of view. Vacations Unlimited needed to notify all the passengers and make sure they got the message. Even with all these efforts, some people still showed up at the scheduled time for canceled trips, which caused a lot of headaches for Vacations Unlimited's people to straighten things out.

With all of this on his mind, Caesar looked as concerned as ever. This was still the first round and according to his intuition, the most critical. The new computer program seemed very nice indeed, but somehow it did not help with the dilemma.

Planning the Next Season and How to be Supported By the Information System—An Analysis

"The new computer program seemed very nice indeed, but somehow it did not help with the dilemma."

What is Caesar's dilemma? All we know is that the nice program claims package D is losing money while Caesar somehow feels it is a good opportunity. That is the dilemma. Can we construct the conflict resolution diagram out of it?

Why should Caesar listen to the calculation of the program? Isn't it because the program is probably based on common knowledge? These are long accepted and agreed upon management norms and models for good decision-making. Using such a normative model is a must for rational and well-grounded decision-making. The rationale of the decisions is of utmost importance to successfully plan the next business season.

And why should Caesar "feel" that the losing package is not a bad one after all? Well, somehow his intuition is different. Is intuition important in the life of an executive? The difference between a brilliant manager and a mediocre manager is the ability to identify a great opportunity as opposed to a common failure. One needs a kind of creativity and inventiveness to come up with such great opportunities that will make the next season especially successful.

So, we can speculate that Caesar's dilemma is the following (Figure 11.1):

This conflict is a generic one. In any particular case of a conflict between an accepted model and one's intuition, the basic assumptions behind the logical links should be checked. TOC challenges several of the most sacred and widely accepted management models. It does not challenge *all* models.

As a matter of fact, there is a clear TOC way to handle the conflict between intuition and the common practice. Verbalizing the assumptions behind the CD' arrow means verbalizing intuition. If you succeed in doing that, you can decide according to rational and well-grounded reasoning, whether they are based on common practice or on new understanding. Certainly it is important to do it in this case.

For Vacations Unlimited, like most businesses, the major constraint is in the market. The success of the exploitation scheme lies in the ability to plan the next season in a profitable way. In this business, the vast majority of the resources they use are external ones. As we shall see, the limitations of hotel

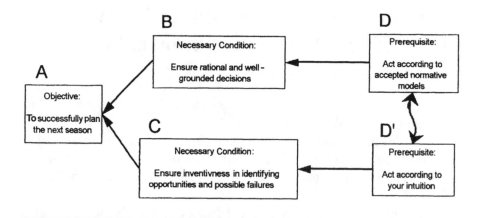

Figure 11.1   A Common Dilemma of High-Level Managers

rooms and/or flight seats may be temporary constraints, but when you look on the business as a whole, the only reasonable constraint is the market, which is difficult to exploit.

What is the Biggest Problem in Planning the Next Season?

There are several problems that come to mind:

1. The need to add all the predicted expenses per package and compare it with the expected price.
2. The impact of the uncertainty. The number of packages to be sold is based on intuitive forecast. Having to plan in September and October for the next July through August season is very uncertain even with the best tuned intuition. Among the unknown factors are the economic situations and the fashionable trends for certain resorts that develop closer to the vacation period.
3. The complexity of matching the various packages to specific flights and hotel rooms.

Which one of the three is the *biggest* one? Let us analyze it from the decision-making point of view.

Problem 1 poses an operational difficulty to gather all the relevant numbers per decision. The method to relate expenses to tour packages needs to be understood and agreed upon.

Problem 2 is the most frightening one. An experienced manager such as Caesar can put forward his own forecast and be aware that he may be wrong. The determination of the reasonable range of the number of packages to be sold is based on intuition. If intuition could be vastly improved, it could help the decision making.

Problem 3 intensifies the previous problem. How to fix the actual flights and hotel rooms. Here, some possible physical constraints may be revealed. In some specific week, we may need more hotel rooms than are physically available. When the rooms are contracted by Vacations Unlimited, the exact dates are not specified. In reality, there might be a contention problem.

The contention problem is much more probable with the flights. If an aircraft of 125 seats is available, the limit of 125 might be binding, meaning the availability of seats is a constraint for that flight. It means that even though there is a demand for 160 packages, we can have only 125. Or we may find it is worthwhile to rent an additional flight for the remaining 31.

When we think of the *other* packages to the same destination, then it is clear that the flight might carry people who have bought different packages. That means, for instance, that if we divide the demand for packages into two different dates/flights, we can fill every flight with 80 people of package A and 45 of package B. The problem is that the duration of stay in Palma de Majorca of the two packages may be different. So, the returning flight that takes back the first round of package A passengers would have to look for other returning passengers, otherwise the flight occupancy will be low. When there are many flights and a large number of packages, the flight planning could be more efficient. Caesar knows that. However, at the generic planning, the detailed allocation of passengers to specific flights seems virtually impossible. Only in the second round is the schedule of the flights planned according to the destinations and packages decided upon in the first round.

Is the Detailed Planning Important for the Gross Decision on What Packages to Sell?

The common management paradigm is to first plan without going into details. Subsequent planning will fill in the details but will not be able to change decisions that were taken at the high-level planning. If the answer to the last question is positive, then we will carefully need to find a way to include the more important details even at the high-level planning.

Let us check using the Palma example to see how the detailed planning might impact the decisions about going to Palma in the first place and then about the package mix. The objective of going to this detailed level is to see how certain details can change the whole high-level picture.

A Simple Numerical Example to Demonstrate the Problems

Let us first assume Caesar's forecast is absolutely accurate! Then we will deal with the problem that it might not be. For the convenience of the reader, here is the table of the packages again (Table 11.1):

Table 11.1   The Financial Analysis of the Packages

Package	Hotel Expenses	Flight Expenses	Other Expenses, Local Expense and Commissions	Total Expenses	Selling Price	Profit per Package	Forecast of Packages Sold
A	600	528	307	1,435	1,795	360	160
B	400	528	92	1,020	1,295	275	200
C	500	528	125	1,153	1,595	442	75
D	550	528	279	1,357	1,300	−57	50

Evaluating the Palma destination, let us refer first to the scenario of having only packages A, B, and C according to the total demand specified in the table. All in all, we have 160 passengers from package A, 200 passengers from B, and 75 passengers from C. A total of 435 people have to be carried by a 125-seat aircraft.

Based on just the forecast and the profit per package we expect to get the following contribution (profits before the fixed costs):

$$160 \times 360 + 200 \times 275 + 75 \times 442 = \$145{,}750$$

For simplicity, assume that all three packages have the same duration. Suppose we first schedule four flights:

Flight 1:   To Palma—55 of A, 45 of B, and 25 of C. Total: 125 passengers.
Back from Palma—empty flight.

Flight 2:   To Palma—55 of A, 45 of B, and 25 of C. Total: 125 passengers.
Back from Palma—55 of A, 45 of B, and 25 of C. Total: 125 passengers.

Flight 3:   To Palma—50 of A, 50 of B, and 25 of C. Total: 125 passengers.
Back from Palma—55 of A, 45 of B, and 25 of C. Total: 125
passengers.
Flight 4:   To Palma—Empty flight.
Back from Palma—50 of A, 50 of B, and 25 of C. Total: 125
passengers.

The distribution of the passengers took into account the inclusion of the whole demand for the C package because it is the best, then the A package, then the B package, for which there are not enough seats to supply all demands. The first flight goes back empty, as no passengers are available to return from the resort. The passengers from the first flight return with the second. The fourth flight goes to Palma empty to bring back the third and last round of passengers.

All in all, there are 160 A passengers, 140 B passengers and 75 C passengers. Sixty people wishing to buy package B are told that there are no more vacancies.

Let us calculate the actual profits, assuming the planning is accurate. First, let us try the profit per package that appears in the table:

$160 \times 360(A) + 140 \times 275(B) + 75 \times 442(C) = \$129,250$ – net contribution of the Palma destination

Let us now calculate the revenue from Palma and the expenses in an alternative way. We first calculate the revenue, then subtract the expenses:

Revenue: $160 \times 1,795 + 140 \times 1,295 + 75 \times 1,595 = \$594,600$

Flight Expenses: $51,000 \times 4 = \$204,000$

Truly variable expenses (hotel rooms + other expenses):
$160 \times (600+307) + 140(400+92) + 75(500+125) = \$260,335$

Profit (the Palma contribution): $588,125 - 204,000 - 260,875 = \$123,250$

First of all, the expectations for the contribution of Palma, which were set at the gross decision level, do not materialize when the four flights are considered. That is because of the constraint of the number of seats in the flights, even though two legs of flights are empty.

And there is another confusion. The net contribution that was calculated from the individual packages sold is different from the calculation based on revenue and expenses. The difference between $129,250 (first calculation) and $123,250 is not so significant to raise doubts about whether one of the methods is really wrong. As a matter of fact, the correct calculation is the second one ($123,250), where the flight revenue was introduced as exact numbers, not based on assumptions regarding the average occupancy of the flights. As an approximation, the first calculation may not seem too bad.

What would happen if we added another flight to accommodate all passengers?

Flight 1:   To Palma—40 of A, 50 of B, and 20 of C. Total: 110 passengers.
            Back from Palma—Empty flight.
Flight 2:   To Palma—40 of A, 50 of B, and 20 of C. Total: 110 passengers.
            Back from Palma—40 of A, 50 of B, and 20 of C. Total: 110 passengers.
Flight 3:   To Palma—40 of A, 50 of B, and 20 of C. Total: 110 passengers.
            Back from Palma—40 of A, 50 of B, and 20 of C. Total: 110 passengers.
Flight 4:   To Palma—40 of A, 50 of B, and 15 of C. Total: 105 passengers.
            Back from Palma—40 of A, 50 of B, and 20 of C. Total: 110 passengers.
Flight 5:   To Palma—Empty flight.
            Back from Palma—40 of A, 50 of B, and 15 of C. Total: 105 passengers.

The expectations are the same as in the gross decision level. Based on the profit per package, it comes to: $160 \times 360 + 200 \times 275 + 75 \times 442 = \$145,750$

The alternative calculations:

Revenue: $160 \times 1,795 + 200 \times 1,295 + 75 \times 1,595 = \$665,825$

Flight Expenses: $51,000 \times 5 = \$255,000$

Variable Expenses: $160 \times (600+307) + 200 \times (400+92) + 75 \times (500+125) = \$290,395$

Total contribution: $665,825 - 255,000 - 290,395 = \$120,430$

The five flights work out to be very close to the four flights. They show considerably less profit/contribution than what was initially expected at the gross level.

However, the difference between the two alternative ways is now much more apparent. The expectations at the planning levels are for about $145,000 net contribution, while the actually realized contribution when the forecast is fully materialized is $120,000. That means something significant is lost in the planning if there is no information support to assign the packages to flights.

The Numerical Example Continues—A Twist in the Planning

Let us check, just for fun, the contribution made when we include package D. This decision is accepted at the highest planning level. As the contribution of every passenger in this package is negative (-57), the total contribution from the 50 passengers will be reduced by 50×57. That means a bad decision at the planning level.

The calculated net contribution is: $145,720 - 50 \times 57 = 142,870$.

We already know that due to the tiny details of the flights, we will have only $120,430. So inclusion of the group should reduce that sum as well when we come to the detailed planning.

Flight 1:   To Palma—40 of A, 40 of B, 20 of C, and 25 of D. Total: 125 passengers.
Back from Palma—Empty flight.

Flight 2:   To Palma—40 of A, 60 of B, and 20 of C. Total: 120 passengers.
Back from Palma—40 of A, 40 of B, 20 of C, and 25 of D. Total: 125 passengers.

Flight 3:   To Palma—40 of A, 40 of B, 20 of C, and 25 of D. Total: 125 passengers.
Back from Palma—40 of A, 60 of B, and 20 of C. Total: 120 passengers.

Flight 4:   To Palma—40 of A, 60 of B, and 15 of C. Total: 115 passengers.
Back from Palma—40 of A, 40 of B, 20 of C, and 25 of D. Total: 125 passengers.

Flight 5:   To Palma—empty flight.
Back from Palma—40 of A, 60 of B, and 15 of C. Total: 115 passengers.

The D passengers were divided into two groups of 25. The first goes with the first round and the second with the third.

The total contribution according to the alternative calculations:

Revenue: $160 \times 1{,}795 + 200 \times 1{,}295 + 75 \times 1{,}595 + 50 \times 1{,}300 = \$730{,}825$

Flight expenses: $51{,}000 \times 5 = \$255{,}000$

Variable expenses: $160 \times (600+307) + 200 \times (400+92) + 75 \times (500+125) + 50 \times (550+279) = \$331{,}845$

Total contribution, package D included: $730{,}825 - 255{,}000 - 331{,}845 = \$143{,}980$

This is very confusing. It seems that the decision to include the package is the right one. How could we know that at the gross decision level it would seem to be a losing product.

Let us summarize the results according to the gross decision:

Decision to go to Palma with just packages A, B, and C. Initial expectations, based on the calculated profit per package, are $145,270. Then in the flight schedule round, four flights will be scheduled, and the net actual contribution will be $123,250.

There is a difference of approximately $22,000.

If the decision to go to Palma with the A, B, C, and D packages is taken, probably based on Caesar's unfounded intuition, then the actual total contribution will be $143,980.

That means that package D is profitable!

Why does the profit per package lie? It is because of the allocation of the flight cost, of course. When package D can be safely added to the same number of flights, it adds to the profits. It does not reduce it.

The simple assumptions about all packages having the same duration provided a very efficient flight schedule. If package A had been for two weeks and package B for only one week, there would have been a much less efficient schedule that would cost much more for the same revenue. What is worse is that in the gross decision level, that effect is very hard to identify. In such a case, the difference between the expected contribution and the actual one will be even more striking.

A TOC Analysis of the Case

What is the throughput of one package? We may treat the hotel room and the other local expenses as truly variable expenses. This is based on the ability of the tour operator to cancel rooms that are not needed. But the real problem is the flights. Every flight carries up to 125 passengers, and it costs the same. So, selling another single package does not necessarily cover the costs of the flight expense. This is similar to using excess capacity.

Instead of calculating a misleading number called "profit per package," we should have related to the throughput per package. The table should have looked like this:

Package	Hotel Expenses	Other Expenses, Local Expenses and Commissions	Selling Price	Throughput per Package	Forecast of Packages Sold
A	600	307	1,795	888	160
B	400	92	1,295	803	200
C	500	125	1,595	970	75
D	550	279	1,300	471	50

From the column of the throughput, it is clear that package D is not losing anything on its own. But because every passenger occupies a seat, the D passengers generate the least throughput. The question is, how many available seats are there.

The planner has the following problem. The desirability of every package depends on its impact on the flight's capacity and schedule. Suppose package C, which generates the highest throughput per unit, is for 10-day vacations, and all the rest are for a week or two, this may create the need to send a special flight that will take only the 75 people who have purchased the C package. How would the C package look then? Much less desirable. In the case of C, the total throughput generated is 970 × 75 = 72,750, which can still compensate for the additional expense of $51,000. If there were only 40 passengers instead of 75, this would be a real loss!

To look for a good enough package mix there is a need to provide a flight loading algorithm. That will enable the planner to look for a more realistic assessment of the throughput vs. the additional expenses of the flights. The current information technology is strong enough to do a finite loading of flights and hotel rooms. It can easily identify if the number of rooms needed on a certain day do not exceed the actual rooms available. Products/packages can be evaluated according to their ability to effectively utilize the seats in the flights—the one resource that is most likely to become a constraint.

The above procedure would be adding packages to a whole mix, checking the availability of seats and rooms, and coming back with the total through-put and operating expenses associated with this product mix.

A computer program built on the planner's needs solves both the first problem and the third. A software program that calculates the profit per package unit may mislead the planner or create the dilemma between the calculation and the intuition, as is evident regarding Caesar.

The above broad guidelines are a replica of the drum-buffer-rope logic of TOC for planning the shop floor. Here, too, you have a master production schedule that is a load on each resource. TOC claims that only very few resources are the constraints. In the tour operator business, flights are the most probable constraint, hence, it should be carefully managed.

If the flights are the constraint, the hotel rooms should have excess capac-ity. That means we should reserve enough rooms so as not to impose a limitation on the use of flight seats.

The question of how many rooms the planner should go for links us to the problem of uncertainty. Human intuition is good enough to catch the "reasonable range" of the variable being faced.

In the example above, suppose Caesar's forecast for the demand for pack-age A had been 120 to 200 passengers. If a computer program assists him in evaluating the various scenarios, it can display the decision factors: which packages to include, how many flights are needed, and what is the expected contribution of the decision. If we suppose that the forecast numbers are just averages, the conclusion will be that the five flights with package D are pretty full and hence, if the actual demand for Palma were larger than the average, some of the demand would not be satisfied. That might lead the planner to reconsider the decision to include the D customers, not because the D pack-age is losing money but because if the actual demand for C or A were larger, much more throughput could be generated providing there are enough seats in the scheduled flight. This is, of course, a gamble between the certain throughput generated by the D package and the possible larger throughput of packages C and A.

The assessment of the reasonable range per package is very important when dealing with the number of hotel rooms needed. Unlike the flights, which take all the packages to Palma, the hotels are matched to the specific packages. Because rooms can be returned by the tour operator, a somewhat more-than-average forecast should be used. The reasonable range can assist in determining the number of rooms to be reserved.

This kind of analysis is necessary to have a real added value of the information system.

A more TOC-oriented planning should deal with the question: What is the system constraint of the tour operator company?

As long as the planning is poorly done, the inability to exploit the market constraint creates a policy constraint in the planning. Once this policy constraint is elevated by better understanding the needs and more appropriate support of an information system, the possibility of an internal constraint emerges. Should the company go to every destination that provides positive throughput? Our intuition says no, but every destination like that adds throughput to the profits of the company and better exploits the market constraint. Our intuition says no because maybe there is a limit to the number of destinations a single tour company can handle.

What determines that limit? The capacity of the planners? The cash the company needs to reserve to pay the airlines and the hotels? The number of professional escorts needed to accompany every package? The number of travel agents? The capacity of the management to organize so many destinations? The number of available aircrafts to be rented for that season?

The crucial thing is to identify that constraint. Once that is clearly identified, the priorities between destinations can be set in a way to maximize the total throughput.

Is such a service organization vastly different from manufacturing?

The Control Aspect

The concept of the cost per package unit has especially devastating effects when it comes to the technique of actual absorption of the costs.

Suppose that when advertising the various packages to Palma, a major mistake was made and package A did not appear in the catalog. Because of that, no package A was sold!

But regarding packages B and C (package D was not included), the forecast had proved itself to be *very* accurate. So, 200 people bought the B package, and 75 bought the C package. Those 275 people occupied the four flights that were scheduled ahead of time. The cost of the four flights was $210,000. Per passenger the cost of the flight would be $763.60. In Table 11.1, the flight costs were evaluated as $528. The added cost of the flight would have given a very low profit of $39.40 for package B! Package C would look nicer but much less than expected.

The conclusion might be to forget package B for next year!

How wrong can your conclusions be? At least as wrong as the above example can show. The actual cost absorption is so misleading because it is not based on cause and effect logic. Hence, a failure of package A has driven package B to look as if were losing. This is how clever people make foolish decisions when they rely on "well-accepted management norms and procedures."

With all the tedious handling of numerical example, this case shows a systematic policy constraint. It seems that high-level planning is more guessing than rational decision making. And the same old cost concepts are also misleading in remote businesses. But the main message is toward the existing information systems. Only rarely are information systems built upon robust managerial thinking. Beware of the evils of wrong paradigms buried in the information system logic!

12 A Crisis at The Small News

This story is about tired people who are the core of an organization. Do you wish to "analyze" the death of a successful organization because some key people got fed up with the daily work? Is it important to understand the personal dilemmas of five people who look for a change in their own lives? This story is still about a business and about management. So, perhaps a management philosophy should say something about it. This book is supposed to be about learning. So, why shouldn't I include topics about which I am debating?

A Crisis at The Small News—Case

Friday morning October 9, Ray Miller sat in his office thinking about what had happened the previous week. He felt a sudden desire to close *The Small News*. Two of his best salespeople had announced their resignations. One of them had written a letter, which was published in *The Daily*, accusing Ray of treating his sales force as slaves and Alan, the manager of graphics department, of failing to effectively support the clients.

The very same week, the trial between The Photo Shop, a small business for cameras, videos, and films, and *The Small News* came to an end. *The Small News* was found guilty of failing to publish a small advertisement by The Photo Shop for two consecutive months. The compensation of $3,500 was not what worried Ray so much. He was ready to pay $10,000 if only The Photo Shop would agree to keep quiet about it. However, The Photo Shop was not really after the money. The company executives were really angry. They had planned their annual big sale for June, and advertising in *The Small News* during the first week of June was part of the planning. They had not anticipated the mismanagement of the advertisement department of *The Small News*.

The Small News had come a very long way since its establishment in 1991. At first it was just a small student paper that tried to treat the problems of the world in a different way—telling the actual news with a twist, trying to present big news as if it were mere gossip, and promoting some local or small incidents as if they were world-class news stories.

Somehow, this treatment seemed in line with the mentality of Mescha, a medium-size university town of some 150,000 people. *The Small News* became a local daily newspaper in 1995. And three years later it is the most popular daily paper in town, with an outstanding circulation of 20,000 copies every day! In 1997, *The Small News* had tried to expand to Segera, the nearest town about 80 miles from Mescha. It did not work. The kind of humor Ray and his friends brought to the paper was not to the liking of the people of Segera.

Ray had established a unique editorial team that could take any news item and give it a humoristic treatment without losing the content of the idea itself. The only exceptions were news about death and crime, which were treated in a dignified sort of a way that only highlighted the mockery of the other news items. The members of the team were Ray; Bridgett, formally the editor of *The Small News* and a sharp redhead who seemed afraid of nothing; Fred, a small shy person who no one believed could be humorous in his writing (Fred also wrote comic strips for TV but never missed a day at *The Small News*); and Chris, the oldest member of the team who used to be Ray's, Bridgett's, and Fred's university professor. Those four people divided the editorial work among themselves, and Bridgett took the global responsibility of editor. Ray participated only half of the time, as all the responsibility of running *The Small News* as a business was on his shoulders.

Advertisers of all sorts found interest in the local newspaper. Ray hired the services of Alan, a retired artist, to handle the creative advertising and graphics department. Alan succeeded in making the advertisements of *The Small News* match the general style of the editorial while keeping the advertising effective. *The Small News* at that stage had some large clients who had purchased fixed space for the whole year of 1998 with just a minor reduction in the price of 12 percent. Apart from those clients, the price for an ad was $20 per inch no matter where and when it appeared within the paper. To keep the right balance between the editorial and the advertisements, only 750 inches were sold, which comprised about one-third of the actual space.

Ray was occupied with his gloomy thoughts when Bridgett, Fred, Chris, and Alan walked in for their daily meeting to prepare the outline for the following day's paper. None of them looked especially happy.

When everybody was sitting at the specially built round table, Fred announced his intention to quit *The Small News* and move to Los Angeles where he had been offered a very good position as a comic strip writer. After the announcement, a long silence fell in the room.

Bridgett was the first to react. "Well, I think we're speaking here about dismantling the whole package. Our real strength lies in the combination of all five of us. I wondered to myself how long we could do it every day. I have three daughters who need their Mom at home. I need some time off. The sad part is that I know that by doing this we will kill *The Small News*—it'll vanish and I don't care what Stanley thinks about it," she said.

Stanley was the only outside investor in *The Small News*. He was the town's richest person. When Ray and his friends were looking for the opportunity to change *The Small News* from a student paper into a regular local paper, Stanley had invested $30,000 to finance the move. He did not argue much when Ray offered him just 25 percent of the shares for it. At the time, Stanley thought *The Small News* was a nice good joke, and he did not mind wasting a mere $30,000 for a gimmick. In the past year, Stanley had shown more and more interest in the paper. He complained that the five managers were taking huge salaries of $250,000 per year, which left the paper with an average profit of only $50,000. He also thought that *The Small News* could sell more ads. Ray had made it clear to Stanley that he ran the show at *The Small News*. He was the first person ever to tell this kind of thing to Stanley. Ray knew Stanley would look for any opportunity to throw him out of the paper and take over the business.

Fred said in an apologetic tone, "By the way, Stanley offered me $1 million for my shares. I assume he did that to all of us."

Bridgett said, "Nice to know. He offered me only $700,000 for the same number of shares. He assumes women's shares are cheaper than men's."

"That is not the point," Ray said. "The question is why none of us notified the others about Stanley looking to buy us out. The offer to me was $1.25 million, and it is not because I'm male, but because Stanley wants me out more than you, Bridgett."

Chris broke in, "I assume it is because all of us have some thoughts about giving up *The Small News*. I received an incredible research opportunity, and in no way could I have done that and kept my work with you guys."

"I agree," Alan said. "The good thing about handling the advertising of *The Small News* was that I got many offers from all over. Only yesterday I got a phone call from *Newsweek*. But, I'm 69 years old, and I don't want to move around that much. I have enough money thanks to *The Small News* to keep me until I die. What I don't like is managing that damn department. I prefer to be busy with the artistic design and leave the administration to some young bright guys."

"Every job has it's mundane aspects. You can't expect everything you do to be great," Ray said angrily.

"Why? That's exactly what *Newsweek* is offering me. Maybe I can take the post and do it from here. I like Mescha," Alan said.

"Are we all determined to leave *The Small News*? Last week was so damned bad that I feel like I want to do something else. I don't know whether Bridgett is right saying that only the combination of all of us can really work. For me, I could try to do the same somewhere else. But, if we sell our shares to Stanley, this is the end of *The Small News* as we want it to be. Stanley will publish many more ads and reduce the criticism of the municipality. He may still have a good business, but I don't want that kind of a business," Ray said.

Bridgett interrupted him. "I understand Fred is determined to move to LA. I know Fred's wife has been longing to move to California. If Fred goes, as a matter of fact if anybody at this table leaves, then for me *The Small News* is no longer the local paper that went all the way to appear on the front page of *The New Yorker* as the most unique local paper in the world," she said. "Believe me, when I do the final editing every day, I can see clearly how all five of us contribute to what *The Small News* really is—something unique. Take out one part, and it all collapses. Fred, you are a good script writer, but there are many like you. I know I can write some funny texts, and I'm a good editor for comic stuff. But there are many like me as well. It is a shame to leave behind something so special like *The Small News* unless we find another unique challenge to keep us moving on."

Chris said, "Yeah, it is interesting how a local paper has become so important for me. I take Bridgett's point that the achievement is probably due to some kind of synergy among us. But if this is true, then *The Small News* will die anyway at some point in time. We're capable people; we can do all sorts of things, separately or together. I wish I could participate in something unique. But what does it mean? Everything I do is to some degree unique. So I want to do something as unique as possible and still make a living out of it. And by the way, unique means something that is positive and significant, too."

"Alright," Ray said. "I suggest that we talk again next Friday and report what each of us is thinking about doing. Right now we must discuss Saturday's paper."

"This was by far the most serious and gloomy talk we have ever had. Suppose *The Small News* is sold to Stanley, how shall we write that news item. 'Five sad clowns bury the fun in the ground' or 'Business takes over the smile,'" Fred said.

They all burst out laughing.

A Crisis at The Small News—An Analysis

What does this story have to do with the Theory of Constraints? *The Small News* is a successful organization that the majority of the owners are tired of and want to do something else. Where should *our* own interests go? Should we analyze the chances that Stanley will be able to make a good business out of it? Are we interested in what each of the five main characters wants to do? Or are we interested in exploring the causes for a possible death of a successful organization?

Suppose we still want to look at the organization and ask the same old question again: What is blocking the organization from doing more?

This question has skipped what should have been the first question. "More" of what? In the formal words of this very basic question: Does *The Small News* have a clear and agreed upon goal?

Hearing the thoughts of five of the owners, we see that there is a difference between the personal goal of the five and Stanley's goal. More, we can suspect that Stanley had different goals from when he invested the money and now when the organization has established itself as a business.

So what is the majority view of the goal of the organization? From what they say, it seems none is especially interested in making more money. Each one is looking for something significant he/she can do. But even that is not enough. They are also looking for a change, either a new challenge, a family interest, or getting rid of mundane work. In other words, they want to do something significant in their life, but each has a different and personal definition of "significance."

So, we have a major problem in defining the goal of the organization, don't we?

Dr. Goldratt says the goal of an organization should be dictated by the owners of the organization. Certainly, the owners have the legal right to dictate the goal. Do they dictate the goal? Different people have different personal goals. Can we say that the goal is a merger of all the personal goals of the owners?

There are several problems here. First of all, if you simply join together the different values of people, you might get an inconsistent sense of organizational value. To show this phenomenon, let us look on three items: A, B, and C. Three people have different priorities in assessing the relative value of the three items:

The first person prefers A to either B or C, and B is preferable to C.

The second person prefers B to either A or C, and C is preferable to A.

The third person prefers C to either A or B, and A is preferable to B.

A democratic vote will establish that for the group of three, A is preferable to B (two votes), B is preferable to C (two votes), and C is preferable to A. The last statement is inconsistent with the other two.

Hence, it does not make sense to combine the personal wishes of a whole group and come up with a precise and consistent set of wishes for the whole organization. Somehow, the merged objectives are turned into something different.

Another problem in constructing the goal of the organization is that it is almost impossible to make daily sound decisions based on a complex set of objectives. As a matter of fact, we realize it is not easy to place decisions being based on making money as a sole objective.

TOC guides us to define the goal of an organization as a single entity to be maximized while all other wishes are made necessary conditions for the existence of the organization. *The Small News* is a good example to test this approach. Should we say that *The Small New's* goal is to provide challenges to five specific people? Somehow I feel uncomfortable with such a statement.

Are there any necessary conditions for any definition of a goal of an organization to hold? The first basic concept of TOC demands that the organization have a goal to achieve. To achieve something, we need not only the goal itself, but also that all the parts of the organization and other stakeholders be clearly aware of that goal. That argument is presented in Figure 12.1.

So, what is the goal of *The Small News*? I claim that for all practical purposes, the goal is the perception of the clients and the employees of the goal. Although that perception is not clearly defined in the story, I assume *The Small News* is perceived as a regular business organization whose goal is to make money. Even if I knew Ray or Fred personally, I would still treat *The Small News* as a business organization while noting that the founders and owners have other personal goals.

I strongly believe that the simplification of the organization's goal to one entity to be maximized and a set of necessary conditions that should be applied up to a certain level are a must for achieving reasonable results vs. the goal. Hence, assuming the goal of *The Small News* is to make money now and in the future, satisfying a certain level of the personal objectives of the five managers in charge seems to be a necessary condition. How much it is absolutely necessary is a matter of judgment. Fred may go, and Bridgett may find out that the group of four is as good as the group of five. In another

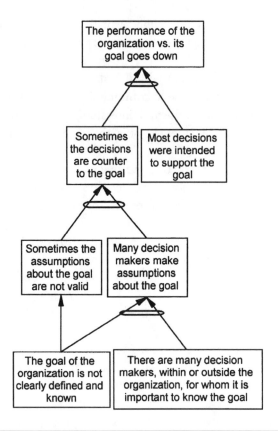

Figure 12.1   Why Do We Need a Clearly Defined and Known Goal?

extreme case, all five will go, and Stanley will bring in others who will keep the business going at least as well. Maybe in such a case *The Small News* will become a different paper. So what? It may matter to certain people, but that is not a necessary condition for the organization to remain with the same kind of a product.

Let us go back now to the crucial question: What blocks the organization from doing more? And more means "money."

The undesired effects told at the very beginning of the story might be caused by Ray and Alan being tired of running the organization and looking for the personal needs of their salespersons. Certainly, we know that all of them are fed up with the daily work.

If we assume that the retirement of each of the five key people may endanger the future of *The Small News*, it is necessary to solve their personal

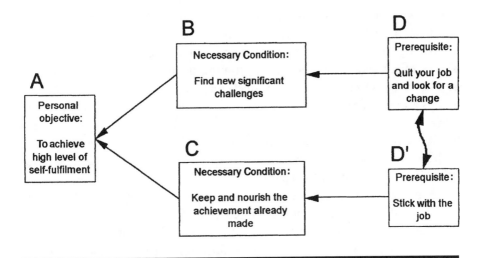

Figure 12.2   The Personal Conflict of the Five, and So Many Others

dilemma. That personal dilemma is presented in the following conflict resolution diagram (Figure 12.2):

The organization strives to resolve the conflict in such a way that the contribution to the organization will be preserved. It may seem that a challenge to the AB or BD logical arrows is preferable than relaxing the assumptions behind the lower part so that each one could quit his/her job and feel content with it.

Is it surprising that certain injections (ideas to resolve/evaporate the conflict) to the CD' arrow might solve the organization's core problem as well? For instance, a basic assumption behind the CD' arrow is that only the daily personal involvement in keeping the past achievement will preserve that achievement. Behind that assumption is another: It is not possible to teach other people to do the delicate job or delivering news with humor.

An idea that challenges a critical assumption behind the CD' arrow is that it is possible to group five to seven young people who, under the supervision of the five founders, will be taught to take up the task of keeping the unique touch of *The Small News*. That may take some time because where the original five managers will have to keep up with the daily involvement. It is a negative branch. The effect of that negative branch can be reduced if we assume that after a short training period, the new management group will step in, but the old group will review the paper every day. Assuming reviewing is less involving than actually writing, the original five will be able to move and start new lives, receiving the written material every day by computer communication or fax and returning their remarks.

Another negative branch from the personal perspective is that selling the shares might put to an end the continuing involvement of all five. Is it a serious negative effect? Maybe. Stanley may be convinced to buy both the shares and the idea that ensures a better transition of the style of *The Small News*.

What will *The Small News* achieve through this injection? A good chance to be at the same state as today but not achieving more because in Mescha there is very little room to expand.

In other words, the former idea solved the immediate threat to *The Small News*, but did not elevate the current constraint.

The question "What blocks the organization from doing more?" has two major impacts. One is the threat to the future. What might block the organization in the near future? That was handled. The other impact looks on what blocks it now, before the threat materializes.

Analyzing the situation of *The Small News* using the TOC's five steps, it seems that the market is the major constraint. There are some problems with the subordination processes. Making money from advertising might be enhanced if the administrative management is able to publish all there is to publish, not permitting an incident such as the one involving The Photo Shop to happen.

A possible internal constraint may be the total space of the published paper. Based on a policy decision, only 750 inches are sold for advertising. Is there a greater market? One can assume that if the market demand was *much* larger, Ray would consider enlarging the paper as a whole. However, enlarging a newspaper by just one or two pages seems technically problematic as most printing machines are built to print eight pages together (then the paper is cut and folded). This case does not provide enough data to further look into this issue.

If there is not enough demand for enlarging the paper, one still might challenge the policy that no more than 33 percent of the whole paper be used for advertising. Is it a precise number? Another claim might be that segmenting the price of advertising according to the location at the paper may yield more money.

All of those considerations are part of the exploitation of the internal constraint. More money can be generated, but once *The Small News* is over the improvements in the exploitation scheme, what else can be done?

Step four of the five steps outlines the link between a strategy and the constraints. Elevating a current system constraint is a must for every strategic planning that tries to achieve better organizational performance in the future.

Simple elevation of the current constraint is not enough for a strategy. Knowing where the organization will be limited in the future is a must for achieving real improvement. The future constraint must be something that makes sense to stabilize the whole organization according to the limitations imposed by it.

What kind of strategic constraint can we think of for *The Small News*? The current major constraint is the market. That is because almost all the citizens of Mescha are captured. A previous try to expand the market to a different city had failed.

When you look for a strategic constraint, you have to consider the unique strengths of the organization. What competence is really unique, so enlarging that it is quite difficult. If that is the case, then the resource of that particular competence should be the constraint of the organization, as the rest of the resources can easily have excess capacity.

My speculation is that the ability of five people to work together in a kind of synergy that results in a unique style is the kind of competence *The Small News* should consider fully using, thus making it a capacity constraint.

Another necessary consideration is to inquire whether there is enough market demand for that unique competence. The failure to expand the style of *The Small News* to Segera may point out that there may not be such a demand.

I believe that every unique demand has universal appeal. By using the word believe, I mean I cannot prove this assertion. I simply feel that if something is "unique," meaning quite a rare competence, there needs to be many potential clients to the unique competence, not necessarily to any particular product that is built upon that competence. Every unique competence can be translated into many different products. The competence of Ray et al, can be used in many products, not necessarily a local paper.

Suppose the potential clients for this kind of humor applied to real world news are distributed geographically. Suppose even that those clients are not interested in having all their news tied up to this humoristic approach but might want to see a special edition of the news done by this approach as part of a nationwide newspaper or TV program.

In outlining the strategy for *The Small News*, expanding beyond the middle-size city it is now a must to fully utilize the precious asset of "The

Five." One of the issues to be checked is whether the tactical plan of recruiting a young group to be trained to run and edit *The Small News* is an appropriate exploitation of the unique asset. Maybe establishing a new national business is a strategic idea that can be done fairly fast and also can provide a new injection to resolve the personal dilemmas by providing a real change and keeping the five in their current jobs as the executives of *The Small News* organization.

Two main messages have been incorporated into this story. One is to establish the organization as a separate entity than the group of people managing the organization, meaning the goal of the organization is not a summation of the key individuals' goals. Still, the motivation of the managers has a great impact on the organization. The second message is to look for the identification of the strategic constraint, as a necessary element in any development of strategy. And a third message, a kind of by-product of this case, is to show again that TOC is relevant in all kinds of organizations.

13 Success Can Be a Problem, Too

T his is the longest story in the book. A very successful company has some problems, and it takes the time to be aware of some of the details. A human behavior consultant is invited by the president to diagnose what is wrong with the company. Is it just the personal problem of the president? From my own experience, I can say that talking to three senior people should give a very good picture of the state of the organization. Of course, not just what they say is important but also what they do not say. If you know what to ask, then you can even spot what the lower-level people think of the management. This is that kind of a story.

Success Can Be a Problem, Too—Case

At the information desk, the white-haired man smiled at Jane. "Oh, you are going to see our president. Aren't you lucky? May I see your ID?" he asked.

Jane frowned. Her name was clearly written at the desk of the man—10 a.m., Ms. Jane Mantle to see R. B. Peterson. She took out her driver's license. The man put the driver's license in a drawer and gave her a visitor's badge with her name clearly printed on it.

"Can I have my license back? I may forget to take it when I come out," Jane said.

"Well, sorry about the inconvenience. We have our procedures here. Anyway, Bill is your escort as long as you're in the building, and it is one of his duties to see that you get your license back. And this desk is manned 24 hours a day," the man said.

Bill, a lean young guy who was sitting there, stood up.

At the elevator, Bill said pleasantly, "It is nice to have female visitors. We don't have many women in the company and not many female visitors."

"How come?" Jane asked.

He shrugged. "I don't know why. I assume there are not many women in this kind of business. I have been here for only three months. I'm looking for a job somewhere else. It is all too strict for me."

Bill suddenly realized he was talking to a complete stranger. "Please, don't tell anyone. Until I find something nicer, I wish to stay here," he said.

"Sure, this is none of my business," Jane said.

Bill escorted Jane to Robert B. Peterson's office. The door was closed. Near the door there was a nice seating corner. There were no secretaries. Bill knocked on the door and spoke into the tiny microphone that stuck out of the door: "Ms. Jane Mantle to see you, sir."

"Please let Ms. Mantle in and wait for her in the seating corner," Bob said.

Bob was looking down the street. He turned around to greet Jane.

"How do you do, Dr. Mantle. I'm Bob. Please sit down. Would you like a drink?"

"No, thank you. And I'm not entitled to a Ph.D., yet. So a simple Jane will do."

"Well, I understand it is just a formality. I took the precaution of asking around about you. You see, we need the right person to help us with our problems. Some high-level executives think very highly of you," Bob said.

"That's nice. I assume you are not used to organizational consultants. So when you do ask for one, it means the problems are perceived as very crucial. I'm honored to have been the chosen one," Jane replied.

"Yes. That's quite right. We're not used to consultants at all. So far, we have done everything ourselves. Even the legal advice is done by people who are full-time employees of E.X.P. It seems to me that when a company needs knowledge about something, it should either send one or more of its employees to a course or hire the right people. Now, this is different. I'm not sure why, but I feel we're facing another type of a problem. I don't know whether you can help us or not. I'm not sure I can properly verbalize the problem. This is a new experience for me. I cannot put my finger on what's wrong—I only know something is very wrong here at E.X.P."

Jane asked, "Can you expand on your feeling that something is wrong? I know that's difficult for an ex-military person, but please try. And don't try to be precise; just let it spill out. What is wrong at E.X.P.?"

"How did you know I served in the army? Did you check on me before coming?"

"No, I didn't," Jane said. "I guessed it. The way you talk carries the characteristics of military service. Also, the profile of the company suggests military people run it."

"Well, that's interesting," Bob said. "I have made an effort to get rid of my old habits. And not all our executives are ex-military people. Never mind. You asked what's wrong. You see, there is a serious debate going now at E.X.P. This is a debate about the future of our enterprise. A very crucial debate indeed."

"There is nothing wrong with a debate about the future of the company. On the contrary, if you had told me there had *not* been any debates, I'd then have said there was something wrong in the company."

"Oh? Are you suggesting that we were wrong in the past for not having debates? Until recently, we pretty well knew where we were heading. During the last six months or so, things have changed. You have to understand that we are a very highly motivated company, and from any business reference we're very successful."

"So, what's the problem? You're successful. That's great. Are you afraid you are going to be less successful in the future? Is it because of the debate going on?" Jane asked.

Bob thought about it for awhile. "You are damn right I'm afraid about the future of the company. I'm one of the four founders of the company. It matters to me very much. The subject matter of the debate is extremely important, but we could succeed either way. I assume that means the debate is not as important as it seems to be. Still, there is something wrong about it. This is an ugly debate."

"Before we continue to talk about the debate I need to know more about the company," Jane said.

"E.X.P. produces listening devices. Small microphones and tiny tapes are our specialty. We also manufacture directional microphones for the same kind of use, devices for recording telephone conversations, and some other equipment of similar technology," Bob said.

"Isn't it illegal to listen to phone conversations?"

"It depends on the circumstances. Our current clients are mainly governments. The police and security agents throughout the world use such devices according to the local law. For such organizations, our products are very important. We are the world leader in those devices. We charge a lot of money for them, and we're so good that we've wiped out all our competitors. We've come up with such superior technology that we've most of the governmental sophisticated listening market to ourselves. Now, this high quality seems to be a problem."

"How come?" Jane asked.

"Our products are so good that their life cycle is quite long, and the clients continue to use them without any urgent need to renew. It seems that our sales volume doesn't increase anymore. As a matter of fact, it has started to decrease. There is no surprise in this. We've anticipated the saturation of the market for some time. We put more and more effort into marketing to get ideas of the client's needs and deliver the message that we can produce and sell the products that satisfy those needs. The pricing issue is also very important—how much is the client willing to pay for the right device to solve the problems? Because of the availability of our high-quality older devices, there is less readiness to pay high prices for improved products."

"Is there an atmosphere of an impending crisis?"

"Not really. We think we can still be profitable for at least the next 15 years. The profits may slowly drop from the extraordinary level of last year to merely good. There is still a big demand for more specialized devices. Although every new specialized product has only a limited market, we still make money. All the new products and the need to find more and more niches in a saturated market have taken their toll on our marketing efforts. We have recruited some excellent marketing people. There are no salesmen. Any contract with a government agency needs the wider scope wisdom for building a winning strategy to materialize such a sale. In my opinion, this is marketing. It includes having access to the right kind of information about the clients. Our development people can do everything that the client wishes provided we know the wishes. Our understanding of the markets still needs to be vastly improved. Five years ago, our super engineers were the stars. Now it is not so straightforward. We have super professionals in every part of the company. We expect quite a lot from every employee. Breakthrough ideas are the key. We expect our people to come up with the sorts of ideas that can bring in more business. Those who cannot cope with that high level do not suit the spirit of E.X.P. Only 10 percent of the people we recruit stay in the company for more than six months. They need to prove their worth in order to stay. We've the best people money can get, and we insist on getting the full value we are looking for from each employee."

"How big is E.X.P.?"

"We have about 1,100 people here at the headquarters. The main R&D and the marketing departments are located here. We have one plant in France and another in Israel. The French facility, of about 500 people, produces most of our products. The subsidiary in Israel has a small R&D department and a small plant for very special products. They have about 200 people. So all in all, there is a total of 1,800 employees. If you take into consideration that our sales last year were about $700 million, you can appreciate the success."

"So, now this debate is going on, and it worries you in spite of this extraordinary success?"

"Yes it does. It is not the only thing, but I don't like it at all. It all started from the business plan I presented at the meeting of the board two years ago. It specified the future drop in sales and profitability. It also stated that we would still make quite a bunch of money for a decade at least. Some of our people didn't like the future I predicted in that document. They have come up with another plan. They want to penetrate the private sector. Undoubtedly, there is a demand for our products there. There are three problems with that. One is that most of the private-sector customers want the devices for illegitimate use. The second problem is that our current clients, the security government agencies, are going to hate the move; they don't want the criminals to use the sophisticated systems they use. The third problem is that this move will pull the prices down. However, I understand the plan and I'm ready to think it over. But the debate is out of control. All of a sudden I'm described as the company's number one enemy. This is a mild expression. Removing the silly old lion is another mild one, and it is presented as a joke. All of these appear on company paper distributed to all the executives. There is no signature, of course. And those documents are coming from various places."

"Are you the only one being treated this way?"

"The campaign is directed against me. The views of my colleagues are split. The alternative plan came from the marketing regions officers. Miles Webster, vice president of North American marketing and sales, supports their scheme. He claims the initiative was the officers', and he simply became a supporter. I believe he is not part of the campaign, but his arguments become more and more extreme. Dr. Yossi Friedman, vice president of special R&D, supports my plan. There are three more vice presidents and four directors who are considered part of top management. Next month the board is going to vote on this. The maddening thing is the way it is being handled. All at the top management level know you are handling the behavioral side of the debate. Go and see them."

"OK. Who is going to arrange the meetings? I see there are no secretaries in E.X.P."

"You're going to see Miles and Yossi today. I took the liberty of setting up the meetings with the two vice presidents. Generally speaking, our managers are very busy and if they don't see any immediate added value in a meeting, they simply don't meet. In your case, you are here because I think this is important. My own special assistant, Don Merril, will meet with you after you finish your meeting with Yossi. You can always reach him on his personal cellular phone, and he is fully authorized to arrange any meeting for you. Bill, will take you to see Miles now?"

That was a clear sign that the first interview was over. Jane stood up, shook hands with Bob, then asked another question.

"It seems that there are very few or no women at E.X.P. at all. Why is this so?" Jane asked.

Bob Peterson looked surprised. "Well, we don't employ secretaries. We use special assistants who are capable of doing much more. The reception desk also handles incoming calls that are not directed to a specific person. I don't think we have any preconception about women at E.X.P.; it simply turned out to be like that."

Bill escorted Jane to Miles. The same type of seating corner was near the closed door.

Miles was on the telephone. He was a younger and taller version of Bob. He also seemed to be even sterner than Bob. He waved for her to take a seat. There was a huge pile of paper in total disarray. He listened for a minute, said, "OK, you go on with it," and hung up. He looked at Jane and smiled:

"Sorry, that call was sort of unexpected. I usually don't allow incoming calls in the morning. They are directed to my voice mail. Most of our external meetings are in the morning and the internal ones in the late evening, and they are not to be interrupted.

So, you are Dr. Mantle. How is the old general these days?"

"Don't you talk with him every day?"

"No. Even before our different vision for E.X.P. surfaced, each of us had too much to do. We meet at the management meetings. He takes a very global view of everything. I need to look at all the sales processes that go through."

"You are doing the real work, and he is just dreaming, doing nothing?"

Miles smiled. "I see you are taking an aggressive approach with me. Bob doesn't really dream. He is not that kind of a visionary. I even like him. I served under him in the army, and he was a great commander. The circumstances that caused his discharge were very unfair to him. It made him want to accomplish very difficult objectives. That kind of determination has been invaluable for E.X.P. I'm not one to belittle Bob's contribution to the company."

"But in your mind, he doesn't contribute to E.X.P. anymore?" Jane asked.

"Dr. Mantle, what is the objective of your work with E.X.P.?"

"The president asked me to identify a cultural problem within E.X.P. and recommend ways to tackle that problem. In his mind, it is related to the debate that's going on. I have not made up my own mind yet. I'm strictly at the diagnostic phase. I can also promise you that I look at the whole E.X.P. management, not the president personally, as my client. My objective is to help the management bring the company into working relationships."

"When you say Bob asked you to identify a cultural problem, do you believe him?"

"I don't have any good reason not to. Why do you think he wanted me in?"

"I am trying to figure it out myself. I appreciate Bob's intelligence. He is trying to maneuver you to back up his position. We at E.X.P. have a lot of respect for professionals. If your professional opinion backs Bob's, he may hope for a victory. I don't think that even your expert opinion will be that powerful to change the course of events. I would like you carefully to refrain from doing that. We're a bunch of fighters here. Don't be the judge. If you operate on behalf of the management, of which I'm a part, don't take any view that may be interpreted as backing one side. This is my advice to you. You have impeccable credentials that you don't want to spoil. "

"So, you, too, have investigated me."

"Of course. You may find out that some of your references had as many as 20 calls from different people at E.X.P. who asked questions about Ms. Jane Mantle and how close she was to receiving her Ph.D. I certainly was impressed by what people think of you."

"Sure, I gave only those who truly love me."

Miles smiled. Jane noticed that the smile was not a friendly smile. Miles was in a detached and carefully controlled state of mind.

"Don't underestimate us. As one of the vice presidents of E.X.P. and one of the four founders of the company, I'm supposed to be very capable. I reached people who don't exactly love you Jane. From the less than favorable things they told me about you, I could conclude that you must be very good in your field. I can even tell you that I think that a certain Professor Manheim is really frightened of you."

"Miles, what is the debate really about?"

"It is about the future markets the company needs to go into. Bob is the only one who thinks of staying only with our current government agencies. If we don't find new markets, E.X.P. will be out of business in less than five years. But where we should enter is still a question. If we identify the right markets, we can prosper for a long time. What really drives the company are two things: making money and being the best in world in what we sell."

A strange buzz was heard. Miles frowned. "Sorry, Jane, this is a little bit unusual. Let me answer the phone for a minute."

He listened for several seconds. "Send Marty with a nice proposal," he said. "He can go as high as $1 million. But it should be clear to Mike that he needs to drop the charges and keep quiet about it. We need to suppress any information about that matter. It should not have happened again."

Miles hung up. He was not as calm as he tried to look. He was silent for awhile. "Dr. Mantle, I suppose you are bound to keep secrets about any issue of your clients," he said.

"Yes, I am. I also signed the secrecy form. What happened?"

"A former marketing person is suing us. He had a heart attack after leading a very large bid and losing it. He claims the inquiry we have made after that failure caused him to have a heart attack. He used to be a very good marketing guy, but he has lost some of the drive. In spite of the heart attack, we didn't fire him. I told him I'm ready to send him to Germany for two years. He refused. He wants to have the special bonuses we give for getting new clients. His income was $200,000 a year, not including the bonuses. Now he wants more money from E.X.P. and to do nothing for that. That kind of greediness is beyond me."

They were both quiet for a minute. Then Miles pulled himself together in order to continue the conversation. "Let me show you what I mean by 'the best in the world in what we sell.'"

Miles pulled out of a drawer a tiny box the size of a lighter. "This is a tape recorder we make. It is a part of the listening devices. It can record for over three hours continuously. What would you expect the quality of the recording to be? Between quite poor and acceptable, right? This morning I brought in a CD of military marches. My own brand of music. I recorded the CD on this tiny tape. Let me connect the tape to the sound system and play it back for you."

The sound of the music was full and impressive.

"No company in the world can do that," Miles said. "This tape can compete with the original CD. And what do we do with it? We sell it as a part of listening devices for police and agencies. They need the quality to analyze the voices and prove who spoke on the recording. Why shouldn't we make a commercial product out of it? This is one-tenth of the weight of any portable tape player and the quality exceeds any portable CD player. There is a huge potential market for it. Why not grab it?"

"Is this part of the debate—going into commercial products?" Jane asked.

"Part of it. This is where I stand. We need to learn more about marketing strategy. We also need to be much better in our deliveries. We have problems in meeting due dates. Sometimes the wrong products are shipped. Every product in isolation functions well. However, if you asked for a small tape and received a large heavy one, you'd complain. In this respect, it is easier to get along with governments than with a distribution chain."

"Do you think that the debate is being handled in the right way?"

"The right way is what should bring us the right decisions."

"And you know what the right decisions are?"

"Yes, I think I do. I think we'll reach an agreement about the global issue. But we need to be more cooperative with each other. Take Yossi Friedman, for instance. I know you are going to see him next. Yossi is a wizard. He can develop a beauty such as this tiny tape. But he won't cooperate with anyone. Can you imagine that he never told anyone that the tape had such precision of sound? I found it by coincident. Our clients don't need that high a level of precision, and if they do I could charge more."

Jane got up from her chair. Miles winced.

"We still have six minutes left. Have you run out of questions?" Miles asked.

"No, but I want to have coffee for some reason."

Miles went to the door and told Bill to take Jane to the cafeteria.

The cafeteria was quite crowded. However, Jane felt it was too quiet. Most of the people sat by themselves reading something. The few who were engaged in conversation sat near each other and talked in low voices.

When Bill brought Jane to meet Yossi, Bill had to enter the lab by himself, and leave Jane outside the lab. He and Yossi came out a minute later. Yossi introduced himself and apologized for not letting Jane into the lab. "Security procedures," Yossi said. "Personally, I think this is exaggerated. Even if you were a technological spy, I doubt whether you would have learned anything valuable. Anyway, let's go to one of the meeting rooms."

Besides his heavy accent and loud voice, Yossi was casually dressed in jeans, sweater and sneakers. He also had a short beard. He was clearly the odd man out in this very formal company.

The meeting room was intended for intimate business meetings. It had four large armchairs and a small table. Drinks and glasses were in one corner, and a coffee machine was in the other.

"I could have had coffee here," Jane laughed.

"I hate American coffee. Whenever I come over, I bring my beloved coffee from Israel, and I have it right here. Would you like to try something strong and different?" Yossi asked.

"I just had a cup in the cafeteria," Jane replied.

"I'll pour you a cup and you decide whether you will take the risk. I need one."

When Yossi sat down, Jane asked, "Isn't it strange for you to be working for an American company?"

Yossi grinned. "It might look even stranger to you to know that I own 29 percent of E.X.P. I'm the largest shareholder. So I don't feel I am working for an American company; I almost feel I own an American company."

"Why aren't you the president of the company?"

"Because the other 71 percent wouldn't vote for me. I'm too strange. I run the Israeli side of the business, and this is what I want to do. I develop the real smart stuff, and most of it is produced in Israel. I want it to be this way."

"You say it as if your colleagues don't want you to do it this way."

"You are right. They don't. They say we are not very disciplined, that the development time takes too long, and that we do not supply on time."

"And what do you think?"

"I think we provide *all* the development needs for E.X.P. The department here is three times as big as mine. Their output is insignificant. What they do mostly is replicate our ideas and claim ownership for them. This is a waste of money. If I were supported to enlarge my department by a mere 15 percent, we could lead E.X.P. into the future."

"How come such a profitable company doesn't let your budget increase by 15 percent?"

"Because we are not good friends. There is a tendency out of mere envy to scale down our achievements. Miles is a good example of that. He is a good bureaucrat. He used to be under Bob's spell for quite some time. Now he feels he is on the way to power."

"Miles was showing me your small tape and its amazingly excellent sound quality."

"And he is certain he can sell it in the consumer market. He is bluffing. Did he tell you how much we charge for it? Five thousand dollars! No one is going to use it for music. Police and secret agents need it for surveillance. If you go to a meeting with some Mafia leaders, they are going to search you. But, this innocent looking box doesn't look like a tape recorder. It looks like a lighter. We also have a tiny transmitting microphone that can be a part of your belt, and no one will really suspect it is a microphone. This *is* worth $5,000. The funny part is that a new competitor has just emerged with a superior product. We are still in business because of our reputation and the huge sales efforts. In a year or so, that small competitor, located in Helsinki,

Finland, may take over the market for that particular product if we don't come back with a better one."

"So, you back up Bob in the debate?"

"We should think about going into the consumer markets, but with different products. That means a lot of new R&D projects. Most of my colleagues don't like *that*. We have become a company in which the salesmen, not the brilliant engineers, are the kings. I agree with Bob that we shouldn't take our current products to the consumer markets. I disagree with him that we should simply stay in our current markets with our current product lines. We can find more markets, and we can broaden our product lines. Anyway, this is *not* the most pressing issue. Bob is shocked that there are jokes about him and how silly he is. Three years ago I went to him with some similar stuff about me. He told me I should take it easy. He said I didn't have enough sense of humor. What about *his* sense of humor. The huge success we've achieved so far threatens to kill us. When Bob; Miles; Patrick, our French vice president of operations; and I left military service, we were hoping for much less immediate success but also for a more promising future than E.X.P. faces now."

After her meeting with Yossi, Jane said to Don Merril, Bob's personal assistant: "For the time being, I've enough material to think over. I get the picture, and I need to build the interim means and actions. I'll call this week with my requests."

Success Can Be a Problem, Too— An Analysis

Jane Mantle has been asked to diagnose the culture of E.X.P. Even from a very brief reading one can conclude that E.X.P. is not a very nice and warm place to work for. So what? Would you refuse to work for such a company? I admit that although I hate certain characteristics of that company, I am not certain I would turn it down because of the money and the challenge.

Is the core problem of E.X.P. a cultural one? If that is the case, we will need to show a significant negative impact of the current culture (after we better define it) on the company's performance. That is quite a problematic claim. E.X.P. is, at the moment, a *very* successful company. It is true that the future poses some difficulties, but we will have to explain how an old cause (I suppose the company's culture is something well-established by now) is going to negatively impact the future.

Another speculation is that the extraordinary success of the company is a (core?) problem. That means it causes negative effects that cause less success of the company. This speculation outlines a cause and effect loop that regulates the intensity of the success. Peter M. Senge, the noted author of *The Fifth Discipline*, had demonstrated similar loops for balancing processes.

There are two ways to determine the core problem of the company. The TOC way is to start with the undesirable effects and speculate about the causes of each, then dive down to the root causes. I use a variation of this and put some key questions as a guidance for constructing the first logical links. The key question is always "What blocks the organization from achieving more?"

In this case, we should add "now as well as in the future." The drop in sales is noticeable now and the concerns for the future are significant.

The other way is to start by speculating about the core problem, then try to see what effects are supported by it, what effects are in conflict with that cause and for what effects we need to find some other cause. This is a quick and dirty method, and it can produce good results. There is one warning: Don't get emotionally attached to the speculated core problem. It might be something else. Anyway, those logical links that look good enough are at least part of the current-reality tree.

As one of the speculations concerns the organizational culture, here is a possible definition of it: The culture of an organization is the set of basic assumptions that are shared by the members of the organization that are perceived as self-evident and dictate the values of the members and guide their behavior.

This definition speaks about "basic assumptions" that cause "values" that cause "behavior." Instead of dealing with the basic assumption, it seems to me we should first look for certain values that are inherent in the story.

From what was openly said, we can deduce the existence of the following values:

- A constant strive to be the best in the world.
- The value of money.
- The need for breakthrough ideas.
- Respect for high-level professionals.
- Everyone must be busy.
- Every individual needs to be an achiever ("we insist on getting the full value we are looking for from each employee").

One may verbalize these values differently of course. What is most important is to generate some of the obvious results from those values. One value that is quite common in other organizations is *not* mentioned. It is the value of cooperation and mutual efforts. When you require personal breakthrough ideas, you get the kind of "individuality" that is characteristic to E.X.P.

What does this set of values produce? I have concentrated on three values (Figure 13.1).

This culture as such leads to an inherent personal conflict. Here is its description as a cloud (conflict resolution diagram) shown in Figure 13.2.

The conflict is between a zero-sum approach and a win-win one. One of the assumptions behind this cloud is that to achieve something of substance for the company, there is a need to cooperate with other people in the company. Another key assumption is that to spread the recognition of excellence, others must be belittled.

This conflict shows a major cause for the undesired effects in E.X.P. I do not suggest any injections at the moment, as the real injections should come from the global company outlook and not from the employees. A common injection from an employee can be to leave the company. Once the objective of the cloud is challenged, the cloud evaporates immediately.

There are several effects in the story that support this conflict, such as the problems between the development group in Israel and that in the U.S. The relationships between three of the founders are also in line with that conflict.

We can assume that right now every employee has found some compromise between the opposing forces. The formal processes also contribute to the existence of enough cooperation throughput the company to be *very*

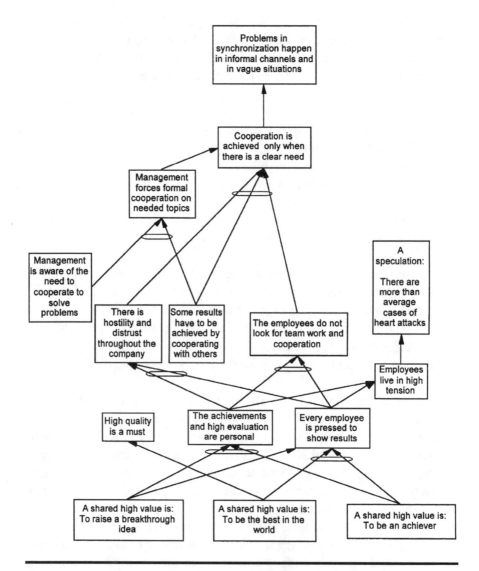

Figure 13.1   Direct Results From Some Cultural Characteristics

successful. Please note that the success has probably generated from the constant strive to achieve.

So, what is wrong? Is the personal rivalry between the top managers and probably among most personnel the core problem of the company? How can we know whether the rivalry is hurting the current and future performance of that company, which is still *very* successful?

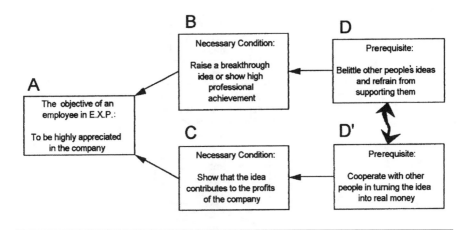

Figure 13.2 The Personal Dilemma of an Employee in E.X.P.

We will need to introduce more undesirable effects that affect E.X.P. We know that the company's market is starting to drop. That is explained by a saturation of the market partially due to the high quality. To expand that market, more customized products are produced. We also know that there are some synchronization problems, and sometimes the customers get the wrong product. Let us try to determine some cause and effect relations between those significant effects found. Figure 13.3 presents another branch of the current-reality tree. The limited market of E.X.P., large as it may be, is going to be saturated. The root cause of the high quality of devices is not something to be changed. The limitation of the market can be changed, and this is going to be checked later.

The top of the tree answers part of the question of whether the problems in synchronization threaten the company. It validates the intuition that it is to be expected that the situation will get worse.

However, the debate within the company is addressing the business problems and looking to go for the commercial market. Regarding that move, we have heard different views. The president stated that he sees a problem in the possible illegitimate use of the products. That may be interpreted in more than one way. It might be explained by a fear of regulations against selling these products in the consumer market. It also may be the case that the objection is based on personal values. It is the latter interpretation that is included in the summation of the company's current reality.

Another factor is included in the summation. The title of the story suggests that success is part of the problem. To check the impact of the success,

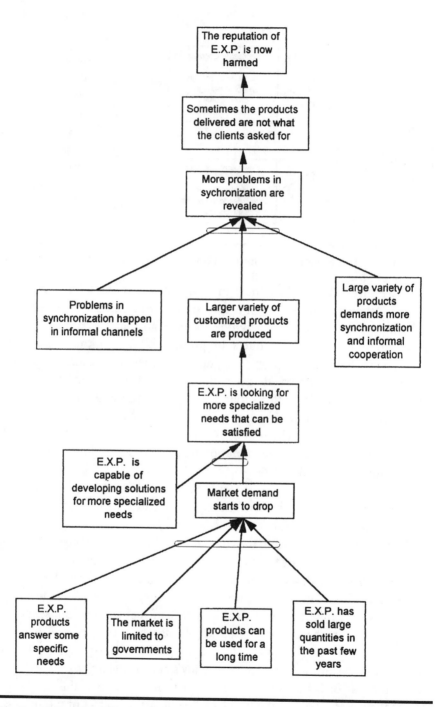

Figure 13.3   The Effects of the Limitations of the Government Market

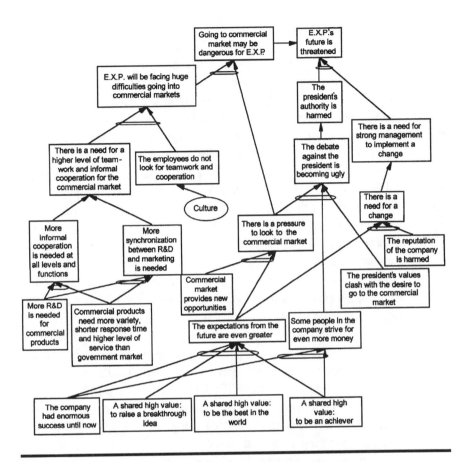

Figure 13.4   The Final Summary of the Current-Reality Tree

it is included in the final analysis. The combination of success and personal ambition leads to very strong pressures that play a key part in the gloomy situation of E.X.P.

The cause and effect links outlined in Figure 13.4 show that the impact of the company's culture is evident. The main problem is the conflict presented by the cloud in Figure 13.2. It can be explained that as long as the company is in the governmental market, the current culture, which is still a constraint, is not that disastrous. It is to be feared that expanding the market is very dangerous unless a change in the internal culture is achieved.

Why is it claimed that the commercial market needs more cooperation and team work than the governmental market? First of all, because the formal requirements of the government contracts allow for formal cooperation to be enough. The marketing people bring the requirements, and engineering does it. In the consumer market, there are no definite requirements. There is a need to predict what consumers like. The reasons for rejecting a product are very wide and difficult to formalize. Hence, a more tight collaboration is needed. Also, high-level technology is not enough. Design is extremely important. So, there are more reasons to team up. These arguments do not appear in the figure. These are more elaborate explanations to the two cause and effect arrows coming out from "Commercial products need more variety..."

As the cloud (conflict resolution diagram) in Figure 13.2 shows, if a challenge to the personal competitiveness is introduced, the conflict might be resolved (evaporated in the TOC terminology). Such a change may resolve the upper part of the conflict. Of course, you may try to challenge the bottom part. A possible challenge like that might be to find more technologies to sell to government agencies.

How can we change a culture? This is the realm of organizational behavior people. Changing the measurements is certainly important. Leadership and self-example are necessary. In any case, such a move *takes time*. It has to start with the top management. For me, Bob Peterson, E.X.P.'s president, is absolutely right in his diagnosis: the ugliness debate is a very worrying symptom. It is not the problem, just a symptom, but maybe it is where the change should start. If the top management agrees on the identification of the problem, the change will start to take place. If the president is right in his business predictions, they still have the time to introduce the cultural change that will enable them to take a change in the business. Sooner or later, E.X.P. should find new markets for their core competence.

The culture of an organization is often the most devastating constraint. Once a certain aspect of the culture is identified as the core problem and the related values are revealed, the problem still cannot be eliminated immediately. That means that for some time the management needs to strive to exploit the current culture while preparing the ground for a new one. The reader may take the story and devise a way to run the organization when the constraint is the reluctance of the people within the organization to cooperate. How do you intend to go into the commercial market? Start now or wait until the new values are rooted in. Please note: If you decide to exploit this constraint by further concentrating on personal achievement, then it is going to be harder to introduce the change of values.

Epilogue

hope that the cases in this book represent real-world management dilemmas. In the real world there are a large number of variables, and some of them cannot be measured at all while others are measured by inaccurate data. It is a daily challenge for the decision maker. I believe TOC is effective in providing a valid way to address those dilemmas.

This *Management Dilemmas* book has caused me to go through a number of dilemmas of my own. Like any other decision maker, I had to make some hard decisions. Like any other experience in life, it is important to compare the initial expectations with the actual outcomes and learn about the causes for any gap that might occur. In trying to express my own expectations for the book, I have had to address the issue of measuring the success or failure of such a book. What is it?

Is it the number of books sold?

Is it how the readers have enjoyed the book?

Is it **for how many readers this book has provoked new thinking?**

Is it how many people now recognize the versatility of TOC in a wide spectrum of environments and problems?

Is it how many people changed their perception about TOC?

Of course, it all ties to my personal goal. As I believe that provoking thought is of the utmost importance to the management world, I am focusing my expectations and hopes on this measure. That brings me to the difficult question: How do I measure whether people were provoked by these cases and the attempt to analyze them? To get some means of measuring this, I enclose my e-mail address so that those of you who wish to argue with me, or just say that this book did cause them to rethink things, can write me about it.

And you, dear reader, what is your next step in learning how to deal with management dilemmas? Certainly, you have your own dilemmas to deal with. If you wish to learn more about TOC, there are a variety of books, and workshops and even computerized challenges that stretch even more the process of learning from (virtual) experience.

Let me bless you, dear reader, with much more thinking, scrutiny, provocation, rethinking, and eventually new insights!

Eli Schragenheim
E-mail address: elyakim@netvision.net.il

Bibliography

The Goal, by Dr. Eli Goldratt, Jeff Cox, North-River Press, Great Barrington, MA, 1992.

The Haystack Syndrome, by Dr. Eli Goldratt, North-River Press, Great Barrington, MA, 1990.

It's Not Luck, by Dr. Eli Goldratt, North-River Press, Great Barrington, MA, 1994.

Critical Chain, by Dr. Eli Goldratt, North-River Press, Great Barrington, MA, 1997.

Goldratt's Theory of Constraints, by H. William Dettmer, ASQC Quality Press, Milwaukee, WI, 1997.

Reengineering Performance Measurement, by Archie Locamy, James F. Cox, Burr Ridge, IL, 1994.

The Theory of Constraints and Its Implications for Management Accounting, by Eric Noreen, Debra Smith, James T. Mackey, North-River Press, Great Barrington, MA, 1995.

Re-engineering the Manufacturing System, Applying the Theory of Constraints, by Robert E. Stein, Marcel Dekker Inc., New York, NY, 1996.

Synchronous Manufacturing, by Dr. M. Michael Umble and Dr. M. L. Srikanth, South-Western Publishing Co., Cincinnati, OH, 1990.

The Constraints Management Handbook, by James F. Cox, Michael S. Spencer, St. Lucie Press, Boca Raton, FL, 1998.

Introduction to the Theory of Constraints, by Thomas B. McMullen, St. Lucie Press, Boca Raton, FL, 1998.

Securing The Future, Strategies for Exponential Growth Using the Theory of Constraints, by Gerald I. Kendall, St. Lucie Press, Boca Raton, FL, 1998.

Project Management in the Fast Lane: Applying the Theory of Constraints, by Robert C. Newbold, St. Lucie Press, Boca Raton, FL, 1998.